**Books are to be returned on or before
the last date below.**

- 5 MAR 2001

LIBREX —

# The
# Business Engineer

# The
# Business Engineer

## Björn L Rosvall

*Teknosell ®*

PETER PEREGRINUS LIMITED

Published by Peter Peregrinus Ltd., London, United Kingdom

Peter Peregrinus Ltd.,
The Institution of Electrical Engineers,
Michael Faraday House,
Six Hills Way, Stevenage,
Herts. SG1 2AY, United Kingdom

**British Library Cataloguing in Publication Data**

A CIP catalogue record for this book
is available from the British Library

**ISBN 0 86341 312 9**

Typeset by Euroset, Alresford, Hampshire
Printed in Great Britain by The Lavenham Press, Suffolk

# Contents

# Part II Development

# Introduction

## Are you a Business Engineer?

**'BUSINESS ENGINEER' is not a job description and you do not necessarily need a degree in technology or engineering to qualify for this honorary title — but it certainly helps.**

This book is aimed at all active people in companies that compete by being the best rather than the cheapest, i.e. companies manufacturing and/or marketing technology- and R&D-intensive products, services or systems.

These companies need Business Engineers in all their key positions.

A Business Engineer is somebody with a very special attitude and a very high competence.

The attitude of a Business Engineer is summarised as aiming to give the customer maximum satisfaction.

The competence of a Business Engineer lies in his or her special ability to explain the merits of our technology applied to the needs of our customer. It is based on extensive training and experience in the industry, plus a very specialised proficiency in business.

**This book aims to provide a short-cut to such a proficiency in business, helping the reader to become a good Business Engineer.**

# How to read this book

In this book we look at many complex issues, but I have tried to make the text relatively straightforward and included many examples from real business life. I have written it in this way to make sure that you can pick it up and read it after a hard day's work — or on a train or 'plane.

The essence of the book is to give you an immediate overview of modern business development. However, if your time is really short, I suggest that you start reading it backwards. First, read the list of contents. Then go straight to the 'Summary of summaries' (Chapter 14). There you will find the main messages of the book concentrated in a few pages. Thereafter, you can read the summaries which appear at the end of each chapter.

To go into more detail, you can decide whether to select those chapters that are most relevant to your own situation, or whether to read the book from cover to cover. If you read a chapter a day (or night) you will have a fresh outlook on business development of a technology intensive firm in only fourteen days. But I stress that I have tried to enable you to 'dip in' as much as you like.

## *Write your own book*

I respect your personal experience and judgement in business and, therefore, I have left some space here and there for you to add your own comments. This way you can, to an extent, customise the book to your situation in your own company.

## *Don't get bogged down in the economics*

This is a book on business strategy. If you know absolutely nothing about Managerial Economics, in a few places you may find the economic discussions to be somewhat too compressed. Don't worry! Such calculations are there as illustrations. They key thing is to get the main messages, which I am sure you will.

# Please meet the hero of our book at OPERATOR Ltd.

Leonard Hewson, who manages a business unit for OPERATOR Ltd., was having a pint with his friend Nick Brown after work. Leonard's mood was foul.

'It is really sad to see,' began Leonard. 'I've been trying to build a position in the marketplace for the last ten years and now that we've got the perfect system for flame control, there isn't enough money to launch it properly!

'All because the financial whizzkids in Head Office have invested our profit from previous years in properties, with loans up to the rooftops! And, as if that isn't enough, they now want to get Charles to pay even higher management fees in order to avoid bankruptcy in the property division!'

'Weren't you about to put your savings into property a year or so back?' asked Nick, smiling at his friend's outburst.

'It seems as if that lot upstairs think I should spend my time reading computer print-outs and explaining the discrepancies from our full-cost calculations,' continued Leonard, disregarding his friend's teasing remark. He was not to be put off. 'There also seems to be some kind of unofficial competition to inflate the calculations so as to make it completely impossible to sell any systems.'

'But how can you disregard costs?' teased Nick again, although Leonard didn't hear him.

'I admire our MD for a lot of things but I'm beginning to suspect he's a bit old-fashioned. During a heated discussion today, I asked him three simple questions:

Do we really have to calculate our prices?

Does every business deal have to bear its own costs?

Do we have to show a profit every year?'

'And what did he say to those then?' asked Nick.

'He didn't answer. He just looked at me as if I'd taken leave of my senses.' Leonard drank deeply from his glass.

## *The heroes of post-industrial business*

The new heroes of post-industrial business have to fight hard.

It's not just that they have to fight competitors throughout the world — even in their home market — they also have to fight the 'wisdom' of the old industrial economy on how to set prices, select business, treat recessions and so on.

My aim with this book is for you and I to find our way to a modern method of how to do business in a post-industrial company.

But what is a 'post-industrial company', you might well ask?

I consider it to be a company marketing technology- and R&D-intensive products, services and systems.

## *OPERATOR Ltd.*

Before we go any further, I would like to introduce you to a (fictional) company I will use as a model for purposes of illustration — OPERATOR Ltd. It is an electronics company producing and marketing flame control systems all over the world; it employs around 200 people.

You do not need to worry about the details of a flame control system, as I am just using it to help illustrate the practical side of a business. Let's just say it is a system that controls flames in combustion processes. I hope using this example will help you see the relevance of the discussions ahead in real-life situations. Where discussion is particular to OPERATOR Ltd., the text or narrative will be shown with a tinted background, to distinguish it from the general text. I hope this style will aid navigation around the book.

OPERATOR Ltd. is ruled with an iron fist by mechanical engineer Charles Victor Armstrong — a strict, dominant Managing Director (MD) whom everyone both loves and hates.

Leonard Hewson, engineer, is now a manager — he looks after the 'Industrial Products' business unit. Officially his title is Marketing Manager (Industrial Products).

Leonard is the former boss of Carol Parker, who now manages the business unit known as 'Consultancy Services'. In Leonard's view, Carol has made her way up the company by convincing Charles Armstrong that there is something special about the marketing of services.

Ronnie Newburg is an old hand. He's been with the company since the start and currently manages the 'Consumer Products' business unit — which makes flame controllers for domestic central heating, OPERATOR Ltd.'s original application.

Nick Brown is the Technical Director.

Malcolm Stone is the Finance Director appointed by the main owner — Investor plc.

Last, but not least, there is engineer Blake Fieldman — Development Manager and minority shareholder in OPERATOR Ltd. He's generally known as 'Brains' and is the genius behind the basic inventions that make OPERATOR successful in the world market.

It isn't necessary, but if you want to know more about OPERATOR before you go on reading, turn to Chapter 12 for additional information about the company's history and position today. You can also find OPERATOR's annual report in the Appendix.

# Foreword

It is evident that the world in which we live is increasingly technological. Today, almost every aspect of our business and social life is affected by technology in one form or another, and this trend can only increase in the future.

Britain has a strong history of good scientific, engineering and technological innovation. Yet there is a general concern that the conversion rate, in terms of the successful commercial application of innovative research and technology, could be much better. The application of science and technology for business success is not solely down to the engineer, but there is little doubt that engineers and engineering play a major role in this process. Certainly, if there are weaknesses in this area it is very much the business of engineers to understand what they are and how they should be addressed.

This book is derived from a Swedish book called *Business Engineering*, jointly published (in Swedish) by the author's company Teknosell® and Exportrådet (the Swedish Export Council) in 1992. Sweden is a small country, but its industry has established an excellent reputation for successful commercial application of technology in new products and processes. The concepts which are captured within the following pages reflect some of that reputation, and will be readily recognised as being sound and meaningful to engineers of all nationalities who have achieved success in this area.

In this English edition, translated and customised for an English-speaking readership, Björn Rosvall offers an introduction to modern business for engineers. With his relaxed style, and novel use of the reader 'eavesdropping' regularly on a fictitious company as an example, the book is undoubtedly easy to read. It does not aim to introduce new thinking or concepts that have not been expressed elsewhere — its strength is the way it communicates much of this new business thinking to engineers who work in the real world.

However, it offers more than training for engineers who wish or need to become more active in their company's business. There is much for senior management and executives too — issues like optimal pricing of technological products and systems, optimal business strategy based on development and profitability, communication of value and benefit to the customer, as well as developmental and structural ideas. All are lucidly covered in a manner that will provoke all readers to think and enable them to obtain something positive, no matter how senior or experienced they may be.

The engineering institutions in Britain, including the newly reformed Engineering Council, are currently active in promoting the role of the engineer and engineering enterprise in the belief that engineering is vital to a successful economy and a better society. September 1996 sees the start of the 'Year of Engineering Success' which has the aim of publicising the positive aspects of our enterprise. Publication of this book, with its inherent belief in and support for the engineer in our business life, is therefore very timely and to be welcomed warmly.

*Dr Alan Rudge CBE FEng FRS*
*Chairman, Engineering Council*

# Acknowledgements

*'Bad poets borrow, good poets steal'*

T.S. Eliot

*In what follows I have tried to be, in this respect anyway, a good poet, taking what I have needed from others and making it shamelessly my own. But such thievery is in great part general and undefined, an almost unconscious process of selection, absorption, and reworking, so after a while one no longer quite knows where one's argument comes from, how much of it is one's own and how much is others'.... To attach specific names to specific passages would be arbitrary or libelous.'*

Clifford Geertz[1]

I have stolen this introductory passage from one of my idols, Hans Johnsson. He uses it in the preface to his book *'Professional communications — for a change'*[2].

Many colleagues and friends have contributed — unknowingly — to the platform of thought that preceded this book. They are so many that I cannot take the risk of mentioning one and forgetting another. I thank all of them for enabling me to be, I hope, 'a good poet', at least in the sense above.

On this platform other friends have helped me to build the present version of the 'Business Engineering' concept. They have done so with very conscious and concrete contributions.

---

[1]Geertz, C., *'Islam observed'*, University of Chicago Press, 1971.
[2]Johnsson, H., *'Professional communications — for a change'*, Prentice-Hall.

I thank my colleague Anders Rehnberg, for helping to write the first draft of this book, assisted by my friends in Hertfordshire. I also thank Robin Mellors-Bourne of Peter Peregrinus Ltd. for his extensive work on the final draft. Last, but not least, I would like to thank all my colleagues in Teknosell®, the international network of business consultants, for their contributions.

Finally, without the inspiration and ideas of Business Engineers in the client companies with whom I have worked over the years, this book could never have been written.

*Gothenburg, 1996*

*Björn Lage Rosvall*

# Part I

# CROSSROADS

Chapter 1

# New times, new heroes

*In this chapter we interpret changes in modern business society
explaining why there is a great need for a new corporate hero (or
heroine, equally) — the BUSINESS ENGINEER.*


## New times

We are living in a new era. In the late eighteenth century, people were
hardly aware that a new era had begun then too. That was the industrial
age, but few realised what it meant! Similarly, today we are not always
clear what the transition to a 'post-industrial' society or age might mean
for us and our business.

Not that the concept of the post-industrial society is new though. It has
been described by distinguished authors like Alvin Toffler[1] and John
Naisbitt[2], both writing in the early 1980s.


### The post-industrial economy

We have many names for the kind of society that will follow the
industrial age or society; the information society, the post-industrial
society, the service society and the knowledge society are just a few
that I have come across.

---

[1]Toffler, A., '*The third wave*', W. Morrow & Co., Inc.
[2]Naisbittt, J., '*Megatrends*', Warner Books Inc.

The essence of this plethora of expressions could well be that, as far as business is concerned at least, we now produce less tangible goods but with more abstract values.

# It is difficult to weigh or measure what we buy

## *The worn and torn jeans*

My mother was born in the early part of this century and I have never been able to explain to her why my son's washed-out jeans, made of a simple workman's fabric and with holes on the knees, cost more than a pair of high quality satin trousers. She thought Levi Strauss was a type of waltz!

What do people really pay for?

## *The invisible telephone exchange*

When I was in the army, I was a signals corporal. I was very impressed by my sergeant who had worked extensively with telephone signalling in his civilian career.

He could walk into any telephone exchange in the country and, after a quick visual inspection, decide the capacity, expressed in the number of telephone lines, and the value of the exchange in pounds. The exchange was built using electro-mechanical components and there was a clear market price per line.

If the sergeant were to step into a digital System X exchange today, it is unlikely that he would be able to say anything about its capacity or value. The electro-mechanics have been replaced by electronics which are now contained in quite a small box. Even worse, he cannot see whether or not the box is operational; he cannot see the most important part, the software, which might not have been installed.

A few decades ago, when a telecommunications authority decided to invest substantially in a new exchange, it used to get lorry-loads of complicated, heavy equipment. Now it receives its operational computer programs as a series of bits and bytes carried on a few disks or even via satellite — and if this isn't followed up by a team of experts to install the system and train the personnel, the system is of little or no use to the customer.

## The seminar that was half as long and twice as expensive

I had an interesting discussion with a corporate training manager some time ago. We were discussing how best to evaluate offers for in-house company seminars.

The normal way to evaluate these is to relate the cost to the number of seminar days and divide this by the number of participants. However, the cost of the seminar is just the tip of the iceberg. The real cost is the money 'lost' in taking key people away from their ordinary tasks which, for a sales engineer, could amount to thousands of pounds per day!

I asked him if, through preparation, skill and a good concept, I taught his people something in half a day that normally takes a whole day, wouldn't that be worth at least twice my normal daily fee? In short I suggested a double fee for half the time.

My friend nodded — he had had the same thought himself. The savings would be substantial. Half a day times twenty delegates equals ten days times thousands of pounds per delegate.

The training manager naturally had a hard time communicating this problem upwards in the organisation. How could he explain that he would pay me 'double for half'?

Perhaps it is my job to help him with that argument.

## Quality instead of mass

The industrial society was characterised by mass production and mass communication. Today, however, it is quality and value for individual buyers that is in focus.

I remember hearing a very interesting definition of quality

$$\text{Quality} = \frac{\text{Function}}{\text{Mass}}$$

As an example, I could mention those clumsy old mechanical calculating machines and how they have developed into today's 'credit card' calculators.

In other examples, the mass has remained constant but has been filled with more and more functions. A carton of milk still contains milk, but today it is pasteurised, homogenised, may have added vitamins and the carton itself can bear information (sometimes even advertising other products).

Nowadays we are selling an increasing amount of invisible values to a more and more narrowly defined segment of the market. Very often we tailor solutions to each customer.

The time is long gone when Ford could say 'you can have any colour you like as long as it's black!' Today customers want to choose between colours, upholstery and accessories, making almost every car unique.

---

## Your comments

*How has demand developed in your industry? Do your products contain an increasing proportion of 'invisible values'?*

# It is easier to steal

## *To steal a punchcard-driven machine*

Do you remember punchcard-driven computers? They existed and functioned until the mid-sixties and in some areas even into the early eighties. They were large, clumsy machines yet could do less than a modern personal computer can do today.

To steal a punchcard computer was an impossibility for an individual.

## *To steal a PC*

Stealing a personal computer is a relatively simple matter, especially if the PC is a 'notebook'. But you would never do that! You are an honest and honourable person.

## *To steal a PC program*

To steal a PC program is even simpler than stealing a PC and — surprisingly enough — many normally honest and honourable people do so.

## *To steal information and ideas*

Information and ideas are the very foundation of the post-industrial or information economy, yet they are frighteningly easy to steal. I know a case where a person in a development department 'stole' over half a million pounds.

The man in question was project leader of a development project and everyone knew this project would require funding to the tune of at least £0.5 million before yielding any income. When it was time to harvest, the project leader resigned, left with a couple of floppy disks and started his own company.

Had this been in the good old days when most development projects required heavy investment in plant and machinery, it wouldn't have been so easy to make off with the value of the development.

## How to protect oneself

There is some question as to whether it is even possible to protect oneself from the theft of ideas and information. I am no expert on the subject so my intention here is just to make you aware of the problem.

Of course there must be something we can do about the theft of ideas, otherwise it would be a great loss to humanity. If there were no chance of remuneration and profit, few businesses would want to invest in expensive development.

The patent laws are still intact but they become less and less efficient as we get more and more base technologies. New applications can be arrived at in many different ways and it is very costly to protect them all with patents. However, taking out a patent is still a viable solution in many cases. Legislation to protect computer programs is also improving, as are methods for copy protection.

Friends of mine built a function into their software which sensed the date; if they weren't paid within the specified time, the program and data were automatically erased.

It is not unusual for modern companies developing software products to spend at least 25% of their development budget on copy protection, such is the importance of this issue. Vivien Irish[1] has recently written a good book, specifically for engineers and technical business people, on the whole subject of intellectual property, to which I refer you for more detail.

---

[1]Irish, V., *'Intellectual property rights for engineers'*, Institution of Electrical Engineers, 1994.

# Your comments

*Of course you have never made an illegal copy of a program but perhaps you know others who have. By how much would the profit in your company go down if half of the key people resigned and started their own company? Is it possible? (Remember, it is not always necessary to have in-house assets and machines. You can often subcontract production. Knowledge and contacts are the crucial ingredients.) How is your company protected against the theft of ideas and information?*

# Investments are becoming softer

## *Investment in fixed assets*

If a company invests £1 million in a machine, we assume that this
expense will be spread over the lifetime of the product. If the lifetime is
five years, we'll add £0.2 million onto the profit-and-loss account as a
cost, and £0.8 million onto the balance sheet as an asset.

## *Investment in product development*

But the visible investment in equipment nowadays is not the end of the
story. Product development often leads to twice the expense of
machines, so in order to develop the products we intend to
manufacture on our £1 million machine, we would need to invest £2
million in product development.

You might think that this investment, which is intimately associated
with the lifespan of the machine, will also be spread over five years. But
is this done in practice? No, these investments are almost always
written off in one year.

## *Investment in market development*

If you invested £1 million in a machine and £2 million in product
development, how much would you need to invest in market
development?

For reasons that we'll come back to, the investments in market
development are often larger still; in our case, perhaps up to £3 million.

We don't even need to ask how this investment is treated in the books
— it's written off in one year!

## *Investment in people*

Hiring a person is generally seen as a cost, not an investment. This
viewpoint is flawed, especially in post-industrial or knowledge

companies (i.e. those that employ many skilled people). The cost of finding the right candidate can easily come to £10,000 or more. But the greatest cost is in training him or her.

The staff used as trainers are often people with high alternative contributions to the business. That is to say, people who play a major role in the business and whose presence is missed when they turn their attentions away from core activities.

So we could be talking about as much as £50,000 investment.

Are you capitalising these investments on the balance sheet? No, they are put straight onto the profit-and-loss account!

## What does good accounting practice say?

Is it right that most of our investments should be written off in the first year? What does good accounting practice say?

The law doesn't prohibit capitalising development investments, assuming that they have a lasting value; so nothing stops us from adding most of our investments onto the balance sheet and by doing so improving our profit-and-loss account.

But companies do not like paying tax. So they gratefully welcome the opportunity for 'excessive write-offs' of development investments, so long as the overall result is still positive.

If you anticipate showing red figures, it could be tempting to capitalise your development costs, but good accounting practice does not allow it very often. The reason is that there is natural scepticism against capitalising half-finished development projects and bold advertising campaigns as it may be very difficult to realise these kinds of assets.

On the other hand, it won't be easy to get the money back on the cost of the machine we have bought if we fail in product and market development.

## What does the stock market say?

The stock market takes a very dim view of companies capitalising their development expenses — this is seen as a desperate measure. It is quite in order, however, to comment that a result was lower because development costs were treated conservatively. This seldom has an influence on the share price, which only goes up when it's clear that the company can actually harvest the profits of its development investments.

What are the consequences of this cavalier treatment of our investments in development?

Mostly it results in a stop/go policy, which means that when we have produced good annual results, we can invest in development. But as development investments are written off in one year, our performance will not be as good the following year, and we may be forced to consolidate or retreat.

You could of course quite rightly claim that if the development investments were spread equally each year, this wouldn't affect the profit-and-loss account. But if that was the case, you wouldn't need to show depreciation at all in your accounts.

The problem is that most investments are not spread equally over the years.

As long as we only have western competitors, these problems have no serious consequences. West Europeans and Americans are equally short-sighted when looking at results, and they are equally prone to alternate frequently between braking and accelerating!

## Consequences with Japanese competitors

If some of our competitors come from Japan, however, the problem gets much worse. The Japanese, in particular, seem to have a different time horizon for the pay-back of development investments, taking a much longer term view. There is a great risk that they will overtake us while we are braking.

## Your comments

*Is your company increasing its investment in development of personnel, markets, products and services? Do you 'capitalise' these investments, i.e. list them on the balance sheet? Are you forced into a stop/go strategy? Do you risk being overtaken by competitors with longer strategic vision?*

# The product life-cycle is becoming shorter

*The pace was slower in the past*

In days gone by, when you invented something, it seemed you had plenty of time to recover your development investments in the domestic markets, before the Japanese or other international competitors found out what you were up to.

When you had enough in your coffers, it was 'time to export'.

*Copying at electronic speed*

When a company like OPERATOR Ltd. introduces a new flame control system, it has two to three years before the Japanese are in the market with a system which is just as good or even better.

If a manager like Leonard Hewson is only given funds sufficient for a half-hearted launch, OPERATOR will not have time to harvest before crippling price competition sets in.

To invest money in the development of new products, services or systems without planning to spend large sums of money on the launch and immediate marketing is the equivalent of saying that a house is complete without its roof!

# The added value increases but flexibility decreases

There are still companies that insist on regarding wages as a variable cost.

Personally I think it's a bigger decision to dismiss a qualified machine operator than an accountancy clerk.

When I joined a manufacturing company ten years ago, it had the usual organisational structure of one person per machine; one working the lathe, one tending the drill and so on. The company-specific training programme wasn't very long, but there was some advantage in this.

If the company was looking to reduce capacity, there was no need to get rid of a whole shift; capacity could be adjusted in line with demand. If it then turned out that an increased capacity was needed, it was no problem to employ a couple more people.

How is it there today?

Most of the production occurs on a robotic, multi-operational machine. If you need to reduce production, you have to get rid of a whole shift. If you then stop the night shift, you lose the production but you don't save on wages because the night shift is almost unstaffed anyway!

The operators responsible for the multi-operational machine have a long, company-specific training programme from the machine supplier in Germany. To dismiss them is wasting capital. In addition it's not very easy to find qualified replacements when you want to increase capacity again.

In many modern companies it is no simple matter to change capacity marginally.

## *The added value tends to increase*

The trend in the post-industrial society is for the hardware or material content of products to decrease and the information content to increase. This means, in general, that the added value increases.

As an example, consider a computer company. The quantity of hardware has gone down dramatically in the last couple of years in relation to the costs of configuration, programming, installation, training etc.

This is most clearly seen in pure service companies, such as computer consultancies, where the value added is sometimes equal to the income.

# Your comments

*How has cost flexibility developed in your company?*

# Distances are shrinking and barriers are disappearing

## *The global village*

'You only live once' I said, and almost bit my tongue as I saw the amused smile on my oriental friend's face. I had just suggested that I was a holy man, one that dared hope for Nirvana after this life.

You could take this as an example of a cultural barrier between myself and my colleague, but it's actually the other way around. Neither of us is particularly religious but we find a great pleasure in comparing the philosophies of the main religions in our respective states.

We both belong, culturally, to a global village which is a growing international brother and sisterhood of people speaking more than one language and travelling like the wind from continent to continent. We understand each other better than peasants from different parts of the same country did in olden days.

Tariff barriers have been torn down. Cultural gaps are being bridged and language obstacles are being overcome. The Cold War is over. Global computer networks span the globe paying no respect to national borders.

These are the words of a true optimist, but even those who have more difficulty in seeing the bright side of things must have noticed that something revolutionary is afoot and this is bound to influence the way we do business.

## *Exporting pit props from Gothenburg*

When the term 'ex-port' was invented, it mainly referred to shipping simple material goods ('hardware') from a port.

A typical example is pit props from the Swedish forests which came to England, via Gothenburg. The goods were inspected, loaded, unloaded, inspected again and the transaction was completed following customary paperwork and, of course, payment. My grandfather earned a meagre living from this trade.

## Selling flame control systems to Pittsburgh

Today, we have an entirely different situation. If you want to sell flame control systems to the foundry industry in the United States, you could smuggle much of the delivery, probably worth twice as much as a ship full of pit props, past customs without much risk of being caught. You could have the important programs and data on a couple of floppy disks in your trouser pocket.

When the volume of business increases, the 'ones' and 'zeros' will probably be sent via satellite, making our deliveries literally 'over the heads' of the customs officials.

By the way, this is not intended to be a lesson on smuggling; rather the aim of these examples is to demonstrate the fundamental difference between international deliveries in the industrial and post-industrial (or information) societies, respectively.

It is natural that this difference influences the way we do business.

Let us consider OPERATOR again.

## Increased local added value

As OPERATOR develops the American market for flame control systems, it becomes more and more untenable to send the bits and bytes from the other side of the world. Staff are relocated to New England.

Leonard Hewson will have marriage problems if he travels over 200 days a year, but without Leonard going over to adjust the systems, they won't work.

So Leonard takes his family and moves to Boston. They all love it there until it's time for the children to start school, then they want to go home.

The fact that Leonard wants to go back is no calamity. It was planned, and now Bob and Jim, recruited locally, can take over.

After another couple of years, the only 'foreigners' in Boston are the Finance Director and two of the board members. The Finance Director, by the way, has just been granted American citizenship.

How much of the added value belongs to Britain and how much can be considered American is up to the parent company, although the tax authorities may want a say in the matter.

## *The advantage of increased local added value*

From the example above, we can see that the higher the knowledge content of our products, the larger the local proportion of the added value becomes. This holds true not just for technical systems but in general, as the demand for local marketing and local service increases.

An advantage with increased local added value is that you become comparable with local competitors for wage levels at least.

Another advantage is that you are affected less by fluctuations in currency, as more of the added value is produced in local currency.

---

**Your comments**

*How does the 'internationalisation' of business affect your company?*

# Time for Business Engineering

*Time for a (r)evolution again!*

The industrial revolution happened some two hundred years ago. According to the well-known American author Tom Peters it is time again. In his book *'Thriving on chaos'*[1], he writes that he believes it is time for a 'necessary' revolution to question everything we thought we knew about leading companies. The basic principle is that new times require flexibility and a will to change, rather than our old fondness for mass production and mass markets, based on a relatively predictable environment (as that environment no longer exists).

I agree totally with Tom Peters.

The reason I put the 'r' in 'revolution' in brackets is that I have a certain fear of revolutions. It can never be a good idea to turn a functioning company upside-down too quickly. A peaceful evolution is preferable.

It is inescapable that the accelerating pace of development has made certain views, inherited from the industrial society, outmoded. For example, in today's post-industrial economy, the following statements can be made, many of which would be considered heresy in the industrial economy:

- 'It is <u>not</u> necessary to show a profit every year';

- 'You should <u>not</u> necessarily export';

- 'You should <u>not</u> calculate your prices';

- '<u>Not</u> every business deal has to carry all its costs';

- 'Salespeople should <u>not</u> be on commission'.

It is time for many companies to change from industrial mass marketing to post-industrial business development, or what I term BUSINESS ENGINEERING.

# A new business method — 'Business Engineering'

Business Engineering is the name I use for a new method for developing post-industrial businesses with technology- and R&D-intensive products, services and systems.

---

[1]Peters, T., *'Thriving on chaos'*, A. A. Knopf Inc.

The word 'engineer' can mean many things. One dictionary defines it as 'to manage skilfully'.

The Business Engineering concept is in stark contrast to mass production and industrial mass marketing. It is based on the fact that you need to create a unique business product (service or system) tailored to each customer.

Business Engineering reverses many of the 'truths' of the industrial economy and often has the opposite view of how to conduct business.

## It is *not* necessary to show a profit every year

If you consider the situation of the post-industrial company, you find it is not realistic to expect, or demand, a smooth and even increase in profits. The post-industrial company is more sensitive to ups and downs in the economy than an old-style company which dealt in standardised base products.

We will develop this statement in more detail, particularly in Chapters 2, 3, 7, 9, 11 and 12.

## You should *not* export

This heading is of course a play on words. What I mean is that it is no longer desirable to sell to as many markets as possible — even if you can sell in Pounds Sterling, price FOB (i.e. charge the customer for delivery), and get away with a five per cent provision or commission to an agent.

It is time to replace 'trickle exports' with 'multi-domestic business development'.

When we have decided on a new geographical market, we should treat it as just another domestic market.

These thoughts are developed in greater detail in Chapters 3, 4 and 12.

## You should *not* calculate your prices

In the industrial society, long product life-cycles made it highly likely that, sooner or later, you would have to use price as your main competitive weapon. It was then logical, right from the outset, to establish the lowest selling price upon which the company could survive.

Nowadays, customers can generally afford to ask for the best, rather than just the cheapest. Who, for example, would buy a three-year old model of a computer, even at a 20% discount?

## *Pricing according to customer-perceived value*

When price is no longer the deciding factor, it is not possible to optimise competitiveness by price calculation. We now have to charge the price our business offering is worth to be able to reinvest in further development to the benefit of both ourself and the customer.

These thoughts are developed fully in Chapters 5, 8 and 9.

## <u>*Not*</u> *every business deal has to carry all its costs*

In Business Engineering we suggest that it is not necessary for an individual deal to carry all its own costs. **We would like it to carry more!**

In the industrial economy, it was natural to have passive pricing as well as passive business selection. Then you knew it was going to be a question of price sooner or later anyway.

I can almost hear you say that cut-throat price competition is not very uncommon today. That is true, but when it is time for a price war, the technology and R&D-intensive post-industrial company will have something new to offer that makes the old cheaper version unattractive to the customer.

When marketing mass-produced commodities it was natural to calculate the prices. Having calculated the lowest price from the beginning, you didn't have to think about business selection. Whoever wanted to pay the calculated price was welcome to buy. In addition, companies could count on having flexible capacity, making the rule that every business deal has to carry its own full costs seem quite natural.

Business Engineering is founded on two conditions requiring us to have active business selection and judge every business opportunity on its own merits.

The first condition is that we have an active pricing strategy based on customer-perceived value rather than the lowest possible calculation of the price.

The second condition is that the majority of our costs are locked in valuable capacity and we cannot change that in the short term.

Our situation is increasingly like that of an airline with one capacity departure per day.

In Business Engineering, we are aiming for active business selection by using systematic opportunity (or alternative) cost calculations, allowing us large variations in profitability compared with our full-cost calculations.

How this is achieved, and what these terms mean, is explained further in Chapters 6 and 7.

## Salespeople should *not* be on commission

The idea that a post-industrial company should not give their sales staff commission individually is not absolute, but the trend is moving in the direction of successful business being the result of teamwork rather than solo performances by the salespeople.

Additionally, commission can cause internal price sensitivity preventing us from true charging according to customer-perceived value.

Please, do not misunderstand me. I believe strongly in remuneration based on performance (i.e. incentive programmes); it is just individual commission on sales that I question.

# A new title — 'Business Engineer'

## *You get the price you can justify*

Pricing according to customer-perceived value is one of the cornerstones of Business Engineering. Once you discard cost calculation as a base for pricing, it is clear that you can charge the price you can justify. Moreover, it is not justifiable to charge higher prices just because you have higher costs.

## The company's new hero

The only person who can justify a price according to customer-perceived value is the one who knows, in detail, what we offer the customer. But that is not sufficient.

He or she also has to know what the competitors are offering, in order for the advantages of our offer to be made clear. But even this is not enough!

The Business Engineer has to know the customer's applications so well that he or she can translate the advantages of our offer into tangible benefits for the customer.

In addition to all of this, the Business Engineer also has to be an able businessperson.

## A technical degree is not essential, but it helps

A Business Engineer does not have to have a technical degree but he or she must understand the customer's problems and how to solve them.

The Business Engineer creates business through knowledge of what the customer needs and what his or her own company can provide in order to satisfy those needs. Of course, working in a technology-intensive company, it certainly helps to have a degree in technology.

## The Business Engineer is the company's most important person

In Chapter 9 we will see an example where the Business Engineer decides whether we are being paid £390,000 or £2 million for a particular system. Is it not true that the person with this choice is one of the company's most important employees?

*The most important task for senior management*

**The most important task for senior management in a post-industrial company is to fill it with Business Engineers and give them the right conditions in which to flourish.**

We will devote Chapter 13 to discussion of this issue.


# Summary

We have moved out of the industrial society and are moving into the post-industrial economy. Distances are shrinking and barriers between people are being torn down.

Changes in the conditions for enterprise are so great that we can now benefit from a shift from industrial mass marketing to post-industrial business development — BUSINESS ENGINEERING.

The emphasis in demand is shifting away from simple concrete commodities towards sophisticated products with an increased service content, making it more difficult for the potential buyer to assess the value of what is being offered. There is no point, for example, in trying to buy computers by the kilo.

I believe the key person in the competitive strategy which I call Business Engineering is the BUSINESS ENGINEER. The unique competence of a Business Engineer lies in an extraordinary ability to explain the merits of our technology applied to the needs of the customer. It is based on extensive training and experience in the particular trade plus a limited but very specialised and qualified training in business itself.

The rest of this book aims to provide the Business Engineer with the theoretical foundation for such a specialised and qualified proficiency in business.

Chapter 2

# Our goal

*In this chapter we discuss the importance of a crystal-clear objective. The most important resource in the post-industrial company is its people. You do not control them to success, you give them the enthusiasm to strive towards clearly stated goals.*

## The importance of a crystal-clear objective

*Not the old 'what is our goal?' discussion again?*

Every author with any self-respect writing about company management is supposed to start by emphasising the importance of goal-setting. I am no exception.

This discussion will, however, lead to something new. It will lead to a concrete starting point for business development and serve as a frame of reference to replace full-cost and contribution calculations as an instrument for profitability assessment.

*The importance of crystal-clear goals, now more than ever*

When my grandfather and his friends — muscles bulging — loaded the ship full of pit props bound for England, any discussion of goals for the work was completely out of order. The idea was to load the ship as quickly as possible and in such a way that the cargo didn't move dangerously during sailing.

When Charles Armstrong, Managing Director of the electronics company OPERATOR, lets his electronic development engineers loose on the world, anything can happen.

'Why the blazes do you have to persist in selling impossibilities in Singapore when you could just as well sell possibilities in Croydon?' he shouts angrily, after having seen the monthly report at the management meeting.

'Because it's much more fun to take Singapore Airlines to Singapore than the 8.14 from Victoria station,' someone could have answered truthfully.

## Management by objectives — the way to stimulate people today

Had my grandfather found one of his men having a good time at work, that person would have received a sharp reprimand, possibly emphasised by a right hook.

Charles Armstrong — Managing Director and Territorial Army officer — has a tendency to react in the same way. The difference is that, when he does, his engineers just spend the time moping behind their desks and don't come up with anything positive.

Following the incident with Singapore, Charles demanded that every trip should be approved by him personally.

Three of the most brilliant designers left and the rest almost drove him crazy by asking him about everything.

So long as people's work can be weighed or measured, you can use direct control and the whip, or possibly carrots such as extra bonuses for shop-floor workers.

Intellectual and creative work cannot be managed that way — you have to manage by objectives. If the objectives are woolly or badly communicated, people will simply do what they like.

This **might** be sufficient, but an individual's objectives do not always coincide with the company's.

*Don't complicate things*

It could be tempting at this stage to start a longer discussion on the theme of global objectives, business ideas, strategy and tactics, but would you do better business as a result?

I think that the following simple advice may suffice:

**Formulate your objectives in the right order — first things first.**

I hope I will make clear how to do this in the rest of this book.

# A company has no objective

If you want to manage by objectives, everyone has to be aware of the company's goals. But were I to go into OPERATOR and ask:

'Tell me, what are the main business objectives of OPERATOR?', I would get a whole range of answers.

'We want to be the world leader in energy savings' — someone might say.

'How do you know that you're the world leader?' could be my reply.

'Well, we've got the best technology of course.'

'Who decides that?'

'I guess the customers do.'

'So if more customers choose you than anyone else, you're world leaders!'

'Yes — you could say that.'

'So you want to be biggest in the world in energy control?'

That the discussion became somewhat confused is not surprising, having put the question as I did. A company itself does not have a goal. A company like OPERATOR Ltd. is just a registration at Companies House.

# Parties interested in the company set the goals

The objectives and direction in business for a company like OPERATOR are set by the parties with an interest in the company (its 'stakeholders').

Who are the interested parties in, say, OPERATOR?

The first group that springs to mind is the **owners**. They have traditionally been seen as having a right to run the company because they have invested money.

But as the substance of the company is becoming predominantly its human resources, the **employees** are playing an increasing role.

**Society** will also indicate the framework within which the company can act.

Without **customers** the company cannot exist.

**Suppliers** are a party you could do without if you did everything yourself, but such a strategy is very dubious.

The problem with the objectives of these interested parties is that they often conflict:

● The employees want to go to, say, Singapore to do interesting work.

● The person representing the owners wants them to go to Croydon to do profitable work.

● Society wants income from tax.

● Customers want low prices.

● Suppliers want high prices.

The company has to accept a compromise between the different parties and their respective aims.

Let us examine the different interested parties in more detail to try to find a common denominator.

## CUSTOMERS: the company's best friend

Explaining why a customer is important to the company is as

illuminating as a description of the importance of water to the Merchant Navy; I will refrain from doing so.

Customers want good suppliers. If we are the best, customers will want us to increase our capacity to serve them.

## THE EMPLOYEES: *money isn't everything*

Most people in our part of the world are spoiled. They have their basic needs secured and therefore self-fulfilment through an interesting job can be worth just as much as the salary itself.

Ambitious people do not like to work in stagnating companies. If your boss is five years older than you, you may well have to wait until you are sixty before you can take his or her place.

Stagnating companies do not open new offices in Singapore. Nor do they install exciting new CAD-CAM systems to keep up with development.

Ambitious people are interested in **growing** companies.

## THE OWNERS AND FINANCIERS: *a profitability percentage is no goal in itself!*

No penetrating psychological analysis is needed to determine the objectives of the capitalist. As a financier, I want a decent return on my invested capital.

However, it would be a mistake to believe that the owners simply want to maximise profitability as a percentage. That is simply a restriction — not a goal in itself.

### *There's plenty of capital available at 20% profitability*

Capital is a very simple commodity with few differentiating features. It is therefore easy to find a well-defined 'market price' for risk capital.

The market price for risk capital varies with risk, base rate of interest and alternative use.

Expressed as profitability ('return on capital employed') the market price varies between 15% and 25%; I suppose the average is around 20% with very little spread (assuming inflation at 5 to 7%, incidentally).

## *20% of one hundred is more than 30% of ten!*

Once you have reached 20% you do not necessarily seek 25%, 30% etc.; you would rather have 20% of a larger amount.

### *Owners want growth*

The conclusion is that the owners want the company to grow with an average profitability of between 15% and 25%. The average has to be calculated over a long enough period; for example, over a full economic cycle or a whole development project.

Unfortunately, few capitalists have the patience to accept this, often resulting in a damaging stop/go strategy.

If you are itching to start a discussion about different ways of expressing profitability, I have to ask you to be patient and wait until you reach Chapters 3 and 4.

## *SUPPLIERS: our second most important resource?*

Many companies miss an opportunity to increase their strength and flexibility if they do not develop an efficient network of subcontractors.

I consider the position of subcontractor to be stronger than being just a supplier; it implies a partnership in which the customer also considers him- or herself to be a supplier with respect to the end user. There is an interdependence.

We should consider all our suppliers as subcontractors; that is to say, partners with a common ambition to fulfil the end user's needs more fully.

Suppliers want good customers to grow. As a consequence, they themselves will automatically grow too.

## *SOCIETY: society wants growth too*

Who wants zero growth?

A connection is commonly made between industrial growth and environmental damage. But today this doesn't have to be the case as much of the growth in a post-industrial society is simply a matter of increased information, not necessarily using up more resources or affecting the environment.

In fact, new technology often demands fewer resources than old; extending the telephone network does not require new copper mines to be opened, instead it is achieved using optical fibres.

Good economic growth can also create new financial resources for improved environmental care.

# The overall objective is satisfactory growth

## *The common denominator for all the interested parties*

We have now found that the lowest common denominator for all these apparently conflicting interests is satisfactory growth.

Customers want good suppliers to increase their capacity; employees want to work in companies where new opportunities are achieved through growth; owners want 20% profitability on a growing capital base; suppliers need our growth so that they in turn may grow; and society itself seeks greater job numbers and opportunities.

## *Expansion with or without hard work?*

There are two ways to grow and increase the turnover of a company. One requires hard work, the other one not. Unfortunately both types are needed to achieve satisfactory growth.

If you are skilled and lucky, you are already operating in a growing market. To grow is then just a matter of not losing customers to the competition. If the market grows fast enough, you can even lose some customers to competition and still grow.

But to achieve satisfactory growth you will usually have to grow faster than the market; you have to work hard to increase your market share.

## *Strategic reasons to go for growth*

We have shown in the discussion of interested parties that satisfactory growth is the overall objective. There are also some other compelling strategic reasons to 'go for growth'.

Generally speaking I cannot become as skilled at doing something every other Tuesday afternoon as someone who does it for a living every day.

Many people believe that economies of scale are more intimately connected with the industrial society than with the information (post-industrial) society. However, this is not the case.

The more development-intensive an industry becomes, the shorter are the life-cycles of the products and the more important it is to have a large market share. It is almost impossible to obtain a good return on development investments with a small market share.

Development warfare generally leads to a reduction in the number of competitors. However, new competitors will appear should the remaining competitors decide to seek excessive profitability increases, provided that free competition exists.

## *Marketing warfare*

The conclusion is that the 'marketing war' will intensify. It is, however, a beneficial war for the customer; competitors try to outdo each other to serve customers with high quality products at competitive prices.

# Summary

The importance of crystal-clear objectives is greater now than ever. The problem is to define a consistent hierarchy of objectives. The names of the different steps in this process are less important than taking the steps in the right order. The starting point for business development is defining the most paramount objective.

A company can only reflect the collected wills and objectives of its stakeholders and interested parties. Of these, the customers are the most influential, thereafter the employees, followed by the owners, the suppliers and society.

That customers are in the strongest position to influence the company has long been obvious. The most recent change in this balance between interested parties is that the employees now have overtaken the owners as the second most important group. This is a result of the shift in emphasis from machinery to people.

Unfortunately, however, the goals and objectives of the parties are still partially in conflict. Customers want low prices, suppliers high; the owners want high efficiency, employees want interesting jobs. But they all have a common denominator.

Customers want good suppliers to increase their capacity; employees want to work in growing companies that offer them new opportunities. Owners want a 20% return on capital but that capital should be growing. Our suppliers need our growth to grow themselves and society wants us to increase the number of jobs to reduce unemployment.

The overall objective should therefore be satisfactory growth, although unrestricted growth can be a recipe for disaster.

In the next chapter you will therefore see why we need financial stability in order to accelerate along the road to success.

Chapter 3

# Will the money last
# all the way?

*In this chapter we will discuss why the balance sheet means even
more than the profit-and-loss account in managing post-industrial
business development. We will also highlight the relationships
between credit, delivery performance and margins and show how to
achieve balanced growth.*

## Unrestricted growth is a recipe for disaster

*The infamous volume sickness*

What Charles Armstrong fears most of all is what he calls volume
sickness. That is to say, his salespeople will accept any price just to get
the order. This should be impossible as they have all heard the gospel
according to Director Armstrong about the virtues of full-cost
calculation!

Full-cost calculations are constructed in OPERATOR to include full
depreciation and alternative interest on the capital employed (in effect a
profit element). The tendency to build in a profit requirement like this
within the cost calculations means that you reach 20% profitability even
when selling 'at cost'.

What you get on top of full cost is definitely not profit, but should be seen as reserves for unforeseen events, according to Charles Armstrong.

So what is the problem? With these rules, all we need to do is sell as much as possible.

'Yes' said Charles. 'But there are those who will try doing business on these highly dubious and dangerous contribution calculations.' He looked sternly about him, ready to come down hard on anyone who dared utter the word 'contribution'. 'And, on top of that, there is constant pressure on the accounts department to calculate prices too low'.

He was visibly agitated as he stood, somewhat imperiously, thinking about all the wrong calculations he has had to correct.

As long as we have no other alternative, Charles Armstrong is right to defend full-cost calculations. An unclear contribution philosophy is irresponsible and a danger to the survival of the company.

Charles is, however, beset by a nagging suspicion that something is wrong. He has found it increasingly difficult to control his design engineers with strict rules and regulations. On the odd occasion he has even had to break his own rules when it was important to land a strategic order, or to fill a half-empty factory.

You will be glad to know that you will have a much more advanced tool for business selection than traditional full-cost and contribution calculations by the time you have read this book. If you do not know what these are anyway, that doesn't matter — you will not have to 'unlearn' them!

In Chapter 7 we will develop a model for alternative-cost calculation which is far better suited to the post-industrial company.

## Balanced and profitable growth is the objective

Having said that the overall objective for the company is satisfactory growth, we must immediately add that to make the growth 'satisfactory', three conditions, or constraints, have to be fulfilled:

1. Growth has to be **balanced**;
2. Growth has to be profitable in the **long term**;
3. Growth has to be **faster than the market's own inherent growth**.

**All three conditions must be satisfied and in this order.**

# International expansion is capital-intensive

*An Austrian surprise for the Finance Director*

Once upon a time a marketing manager decided to attack the Austrian market rather than penetrate his home market further. Perhaps this was the correct decision; possibly he had taken whatever market shares were available in his established markets, so now it was time to attack. Or perhaps it was for an all too common non-strategic (human) reason, or simply that the grass always seems greener on the other side of the fence! (We will develop this theme further in the next chapter.)

Whatever the reason, the marketing manager sat in an old coffee house discussing business matters with his new Austrian friend, a Viennese distributor. It quickly became obvious that the British payment terms of 30 days net were totally foreign to the Austrian. The distributor could show that all other suppliers gave him 90 days. That was surprise number one.

No sooner had the British marketing manager reluctantly agreed to these payment terms when along came surprise number two. The distributor didn't want to hold any stock!

'But, my dear friend...' the Austrian began, 'I don't know your product range. Initially, you'll have to guide me. I'll give you a very generous offer,' he continued; 'you can use my warehouse completely free of charge. If you send me the products you consider best for my market, I'll tell you when I've sold them and you can invoice me.'

This way of doing business, on a consignment basis, is a nineteenth century model. Then, interest rates were low and there was little or no risk of obsolescence. Those conditions do not apply today and therefore I believe this type of business is no longer viable. The 'generous' offer of lending the warehouse is peanuts in this context, but that was not the worst part of the deal.

The main drawback was that the distributor was under no pressure to sell the stock. Every product he could convince his supplier to put into stock represented a potential contribution of perhaps 30% for the distributor.

But the marketing manager was keen to get started and accepted the proposition. He didn't yet know how to get out of the situation. He should have said:

'My dear friend, how long do you think it will take you to sell a product from stock? More than six months?'

To this question, the distributor, if he wanted to appear competent, would have to say no.

'Good,' the rejoinder might be. 'Let's agree that you get an initial 90 days credit for this first order — that means you don't have to pay us until you've sold the goods.'

But, as I said, he hadn't learned that yet so he had to manufacture goods for stock that didn't turn over more than twice a year.

The end result of this negotiation was that, a year or so later, the Finance Director pulled his hair and sighed:

'We're selling and selling, but I have to keep borrowing more and more to finance it.'

## *Marketing assets — the biggest investment?*

It never ceases to amaze me that a board, without discussion, often allows a marketing manager to invest millions in marketing assets such as credit and stock, while the production manager is being put through the third degree in order to invest a couple of hundred thousand in equipment.

Well, back to Austria. What were the financial consequences?

|  | Million Schilling |
|---|---|
| Austrian turnover in first year | 100 |
| A quarter of this, 3 months out of 12 in a year, was tied up in credit | 25 |
| Six months' sales (stock turn rate of 2 per year) was tied up in stock, which would have been 50 million if the value of the stock was equal to the sales value, but the accountants stated that 60% of 50 million = was a more realistic inventory value | 30 |
| The Austrian business therefore required an investment in marketing assets of 25 + 30 = | 55 |
| Fortunately, the capital employed, the working capital, didn't increase by that amount. Just as the company had to give its customers free credit, they themselves got free credit from their suppliers. There are also other 'free loans' such as VAT, advance payments from customers etc. | |
| In total, these free loans amounted to 10% of sales | −10 |
| So this business meant a net increase of 55 - 10 = in working capital, i.e. £1.6 million at the time. | 45 |

If our owners have requested a minimum profitability of 20%, what minimum profitability does this translate into?

The answer to this exercise in calculating percentages is £320,000 (i.e. 20% of 1.6 million).

If we continue the arithmetic and ask:

'What is the lowest net margin, as a percentage of sales (100 M Schilling = £3.5 M approximately), that just covers the requirement of return of the additional capital?'

The answer is 0.32 M / 3.5 M = 9%.

This percentage could be called the 'opportunity cost' or the 'alternative cost' of capital tied up in marketing assets. I prefer to call it **sales interest** for short.

It is important not to bury the capital costs of your marketing assets in product calculations, but instead highlight them as an important planning and negotiating factor. That's why I've given them the name sales 'interest'. You are free to choose another. The main point is that you accept my line of reasoning.

By the way, it is very unfortunate that terminology in business economics is not fully standardised. Every company develops its own language. The key thing is not to try to memorise the expressions I use in this book but to concentrate on the logic.

For example, the expression 'alternative rate of interest' is sometimes called 'internal interest'. The logic behind this is that the board tells the management team that if they cannot achieve a return on capital employed of X% (in OPERATOR's case 20%), then the board has better alternatives for this capital.

Now let's return to Austria for the last time.

The Austrian business had a profit margin of 15%. The 'sales interest' was 9%. So nearly two thirds of the net profit margin was spent in paying the alternative interest on the capital tied up in Austria. If you consider the risk of obsolescence that was built up in the Austrian distributor's stock, it is questionable as to whether we had any profit at all in Austria.

## *Should we be a bank for our customers?*

'I refuse to be a bank for our customers!' exclaimed Charles Armstrong, when one of his sales engineers was bold enough to propose a few months extra credit to win an order. 'If they can't get credit with their bank, there's no reason for us to take the risk of lending them money!'

An MD like Charles Armstrong has every reason to hesitate to use credit as a competitive weapon. It can be expensive, especially if your own financial resources are strained.

But if you do have good financial strength, you could consider using credit like any other competitive weapon. What does a month's extra credit cost? And what effect does a more generous credit policy have on our market share? Let's consider these questions one at a time.

What does an extra month's credit cost? It costs at least one twelfth of the alternative interest. If the alternative interest is 20%, the cost is about 2% of the sales value — if it just was a matter of the two per cent, an MD like Charles Armstrong would hardly make such protestations. But he is afraid that the customer won't pay at the end.   '... and then the cost is one hundred per cent!' he would growl. Of course this is only true if the customer is genuinely expected to go bankrupt.

One way of estimating the probability of whether a customer will go bankrupt is to express our costs for bad debts as a percentage of sales. If we normally lose half of one per cent of our turnover through bad debts, and this customer is considered to be a ten times higher risk than the average, you can actually quantify it as an extra credit risk of 5% of the total amount.

> 'Well, if they can't sort out their credit problems through their own bank, we should consider them a credit risk too', insisted Charles Armstrong.

You could of course argue that the bank would not take greater risks than those they are paid for, and their interest is hopefully lower than the margin we would expect to get on the business.

This way credit can be seen as an alternative to other competitive actions. Adding the alternative rate of interest (2%) and the risk element (5%) gives us 7%. Obviously 7% of the sales value is significant, but it is less than a price reduction of 10%. On the other hand, you could afford to reduce delivery times considerably if you used the money to keep a greater stock of the product instead.

Overall we have to agree with the likes of Charles Armstrong that an indiscriminate extension of credit times is a tremendous waste of money.

## Delivery service or obsolescence

Nowadays the idea tends to be not to have stock, because of the high alternative interest but also the risk of obsolescence. This is some of the justification for 'just-in-time' (JIT) business and manufacturing.

But there are many good reasons for keeping stock, the least common of which is that you might want to exploit a special offer.

The most common is that customers cannot wait for deliveries and that competition forces us to offer a short delivery time. Therefore we have to guess when and how much the customers want to buy.

If we guess wrong we are left with high product and/or raw material stock and unused capacity. Shouldn't we then use our capacity to add value to our stock and produce more products and systems?

In this situation a company is often torn between the accountants' desire to minimise stock value and, alternatively, high capacity utilisation.

If you have financial resources and the continued processing does not reduce the value of the stock, and if the alternative value of the capacity is zero, it has to be sensible to run up the stock, provided and until you can sell it.

The cost of running up the stock progressively for six months is 20% first divided by 2 (20% is for a whole year, and we are talking about half a year), then again divided by 2 (we start at a level indexed as zero and end at index 100; assuming a straight line build-up the mean will be 50, i.e. 1/2) which is 5% — but of which amount?

If the alternative value of the capacity is zero you have to calculate it on the purchase value of the material and components. If this is 20% of the sales value, the additional stock interest is 1% — comparable in cost to a 1% discount.

The interest cost of buying time is therefore not particularly high. What is more expensive, however, is the risk of obsolescence.

If we are not absolutely sure that the stock we're running up is going to be sold in a reasonable period of time, we could eventually be hit by large losses if we have to sell off the stock at knock-down prices.

One way of avoiding this problem is to increase the stock only when there are confirmed orders; that is to pre-produce.

An acquaintance from a shipyard told me a very interesting and educational story. It is about keeping calm when committing to delivery times.

He claimed that most salesmen give away their best delivery time quite unnecessarily. When asked how soon he can deliver, he always answered with the question:

'How soon do you need it?'

Say they replied 'June at the latest'.

'Impossible!' he would state. 'My earliest delivery date is December.'

'If you can't deliver in October you can't have the order.'

'OK, let's say November' he would concede, knowing full well he could have made the delivery in May.

Using this method, he kept his capacity free in case he came across another, more urgent order. If not, he could start the November order early.

Does that give you food for thought? (Obviously it should not be pushed to extremes and actually deter the customer.) It is particularly useful for service companies; they often believe they do not have any stock.

## Do service companies really have stock?

Many believe service companies do not have stock.

From a financial perspective you can define stock as 'resources used but not yet invoiced'. With this definition, work-in-progress is definitely stock. Some service companies are not aware of this, due largely to inefficient accounting for time spent on a project.

In a consultancy company with long projects, the result in any given year is — to a large extent — dependent on how work-in-progress is valued.

There is a cynical story about how to be certain to make a brilliant career as the MD of a consultancy. You have to be appointed towards the end of the year. Then that year's figures will be considered your predecessor's. When you go through work-in-progress before finalising the annual accounts, make sure you write it off, under the pretext that 'it probably won't amount to much anyway.' This will affect the current year's results negatively but that is not your problem. When you invoice the completed projects the next year, your first annual result can be brilliant. This will particularly be so if you now take the opportunity to value all work-in-progress as highly as you possibly can. It is then advisable to jump ship in time to avoid responsibility for the following year's results. You can then say afterwards: 'look what happened when I left the company'.

This method, by the way, is equally applicable to manufacturing companies. The story simply illustrates the impact of stock valuation on the result of a company selling services or sophisticated systems.

But the fact remains — service companies do have stock, or work-in-progress if that is what you prefer to call it.

This has to be financed. To have stock can be advantageous if it is the result of strategic production to increase capacity utilisation, as in the shipyard anecdote I mentioned above.

## *Capital tied up invisibly*

The balance sheet has never been an uncontroversial way of representing the financial position of a company. Sometimes it is said that the only indisputable items are the debts and the cash in the bank.

Debtors can be questionable and stocks can be obsolete. Increasingly rapid technical developments can halve the value of machines and fixtures in just one or two years.

Fixed assets have hitherto been the foundation of the balance sheet and still are, even when many companies have become painfully aware that a property isn't worth more than what someone is prepared to pay for it.

To add to traditional uncertainty, we are now starting to find that an increasing proportion of a company's assets are no longer entered at all. As discussed in Chapter 1, substantial investments made in business development are written off directly in the profit-and-loss account.

The more a company becomes a knowledge-intensive company, the more undervalued the balance sheet is likely to be.

## The impact of marketing mix on tied up capital

Early in its history, OPERATOR almost went bankrupt in spite of rapidly growing international sales at reasonable margins.

When OPERATOR made its first attempt to sell into North America it met with a very enthusiastic agent in the USA.

'Your only problem will be production capacity!' he boasted. 'You have to guarantee me an even flow of equipment over the Atlantic, as delivery time is crucial. No more than eight weeks is acceptable in this market! And one more thing — you'll have to adjust your terms of payment to the usual 90 days.'

The agent sold a lot of systems, but not in 'an even flow', instead more of the 'ketchup bottle syndrome' — i.e. nothing, nothing, nothing and suddenly a whole lot all at once — and seldom systems that had components or subsystems in stock, but adjusted applications that had to be produced as rush orders.

One day the Finance Director summoned a crisis meeting. He opened by telling the Marketing Manager: 'The more you sell, the more I have to borrow! Now the bank says we have reached our limit!'

'So you don't want me to sell any more systems?' Leonard Hewson asked in disbelief.

'Sure, but not on the present terms! Look here!' the Finance Director answered, putting a transparency onto the overhead projector:

|  | **% of yearly sales** |
|---|---|
| **Accounts Receivable** | |
| 100 days credit/365 days | 27% |
| | |
| **Inventory** | |
| 180 days/365 days (2 times/year) = | 50% |
| evaluated at 60 % of sales value = | 30% |
| | |
| **Free loans** | −5% |
| | |
| | 52% |

'For every million dollars of sales I have to finance half a million!'

'But our terms of payment are 90 days — not 100!' objected Leonard.

'Yes, but there is always enough overdue to make it 100 in reality!'

'What are free loans?'

'Accounts payable, sales tax etc.'

'Can't we increase those?'

'You will definitely need to have more of those. You shall have to demand some help from the customers to finance this business — from now on the terms will be 50% upon contract, 40% after installation and the final 10% after 90 days. And one more thing: no ordering of components or subsystems without a signed order from the customer!'

Leonard broke into a rage over this proposition and accused the Finance Director of 'ivory tower dictatorship a safe distance from the customer'.

'How can you expect me to sell any systems with those conditions?'

Nick Brown, the Technical Director, broke in: 'Come on Leonard! The way you and the agent bow to the customer's special wishes, we only make one or two of each model anyway. Why not present our products as 'customised systems'? Then nobody can demand that you ship from stock! Custom-made systems are always part-financed by the customer.'

Leonard took to the idea. Instead of selling standard systems with a multitude of options, they changed business philosophy and promised the client a customised solution, provided he could wait for delivery. It now became quite proper and natural to ask the customer to help finance the order, as this was to be custom-made for him.

Through better attention to the customer's situation, the margins could be improved and the expansion could be self-financing.

Next year the Finance Director could show the following transparency:

|  | % of yearly sales |
|---|---|
| **Accounts Receivable** | |
| 10% of sales at 100 days credit/365 days | 3% |
| **Inventory** | |
| 30 days/365 days (12 times/year) | 8% |
| evaluated at 60 % of sales value = | 5% |
| **Free loans** | |
| Accounts payable, VAT, etc. | –5% |
| Advances from customers  (4/12)x50%= | –17% |
| | –14% |

'Now you can go on selling. The more you sell, the more money I can put in the bank,' beamed the Finance Director.

'Why do you cut down the advances to 4/12?' asked Leonard.

'That's the average project time,' answered the Finance Director. 'Ideally I'd like to get the full payment a whole year in advance. But if you ask for a year to deliver, even I understand that you won't get any orders.'

## Your comments

*What percentage of turnover do credits and stock represent in your company?*

# Is our dependency on the banks healthy?

*'Why don't we borrow?'*

Charles Armstrong has a failsafe way of cutting short any grand plans with a stern:

'We haven't got any money.'

Woe betide the person who happens to say:

'Why don't we borrow?'

He or she will be given a lecture on the virtues of frugality they won't forget.

Charles Armstrong doesn't realise it but, subconsciously, he considers the company to be his own. Most of all he'd like the company to be his without any loans. His great disappointment in life was when the heirs of the founder decided to sell OPERATOR Ltd. to an investment company rather than to hand it over to him.

However, if he cannot be independent of the shareholders at least he wants a certain degree of independence from the banks. That's why he wants OPERATOR to be a financially sound company.

## *How strong do we need to be financially?*

Financial strength can be defined as the company's risk capital in relation to its total assets. How financially strong we should be is determined by the risks our company takes in the course of its business.

There are two good reasons not to aim for too high a financial strength.

The first is the lever effect, where the greater your financial strength, the lower will be the profitability on your risk capital.

Suppose you have a total working capital of, say (an index of), 100 and your profit is 20.

If you have borrowed 80 out of that 100 at 10% interest, the bank takes 8 out of the 20 you have made, leaving you with a profit of 12 on your own capital (which is 20). Your Return on Equity will therefore be 12/20 or 60%.

If you had borrowed 50 out of 100 at 10%, the bank would have taken 5 out of the 20 you made, leaving you with 15 on your own capital of 50. Your Return on Equity would have been 15/50 or 30%.

It is good to keep in mind that the lever effect works both ways, as some highly indebted property companies have learned recently.

The second reason not to be too strong is that an unnecessarily high level will restrict our expansion, which may lead to loss of market share. This, in my opinion, is the main reason to consider carefully just how financially strong we want the company to be.

As a rule of thumb, the average company should be satisfied with a financial strength of 30%; that is to say, with risk capital equal to 30% of total assets. This can be a starting point when discussing the level in one's own company.

## Your comments

*Consider your own company. If your income seems stable and easy to predict, and if your costs are flexible, you can require less than 30% equity ratio. If your income fluctuates more and is more difficult to forecast, while at the same time your costs are harder to adjust, you should require more than 30%.*

## *Estimating financial strength*

The only undisputed figures in the balance sheet are the cash in the bank and the debts. Risk capital, and thereby financial strength, is dependent on the accuracy of assessment of the assets. Not even assessment according to replacement value is indisputable, as every greengrocer knows.

The problem with many companies today is not so much that there is 'air' in the balance sheet but rather that the visible financial strength is too low.

As the hardware or material content of products decreases and investments become softer, so the visible risk capital — or equity — becomes smaller. We tend only to show development investments on the balance sheet if they result in hardware (i.e. concrete physical things).

It is easy to understand the reasons for a pure 'knowledge company' trying to build up substance through property and share portfolios.

There is an interesting discussion going on about re-evaluating the assets of knowledge companies in their balance sheets. Words such as 'human capital' are beginning to be heard.

When we want to develop our business, however, it is not what the analysts say about our financial strength that counts, so much as the opinion of the bank when we want to borrow money.

I once took part in a round-table discussion arranged by one of our main banks, discussing future market development. We concluded that traditional markets for banks were decreasing which, naturally, was worrying.

The traditional way for banks to do business is to lend money to traditional companies that can show collateral in traditional assets. The discussion took an interesting turn when a banker managing a branch close to one of the universities said:

'What do I tell that engineer with the PhD when he comes in asking for a loan, when the only collateral is a patent and licence agreements in ten countries?'

The question is by no means new but it is becoming more common; traditional capital investments are declining as a percentage of total investments.

Even banks have become more used to reading the 'invisible' parts of the balance sheet, but their basic business concept is nevertheless poorly geared to the modern post-industrial or knowledge company.

The financial market will no doubt catch up with the development towards invisible assets, but in the meantime there is still great risk in being a post-industrial company without sound traditional financial strength.

## Your comments

*Try to view your company's balance sheet through the bank manager's eyes. You can assume that the management, when visiting the bank, has given an excellent account of the company. By how much would you extend the overdraft?*

## *We'll have to rely on our parent!*

If you are part of a larger concern, you normally rely on the parent company to provide financial strength. Calculations of your own financial strength might therefore feel less important. You might not even have your own balance sheet.

However, I believe it is a questionable practice to remove the financial responsibility from subsidiaries, divisions and business units.

I can understand the thinking behind concentrating financial resources 'at the top', but that does not prevent other companies and business units in the group from having independent balance sheets. You can, for example, assign a share capital to them and give them an overdraft facility with the parent company.

As investments become softer, it becomes more and more difficult to answer the question 'Is this expenditure a cost this year or an investment for the future?' Expenditure that traditionally has been 'cost' on the profit-and-loss account, and therefore affected that period's profitability, should in some cases be redefined as 'investment' and apportioned to the balance sheet.

As the result for a particular period is more and more dependent on the right costs and income being apportioned to that period, it is important that these are calculated correctly. But who should do it? Often the only person who really knows what resources have been used to produce a result is the front-line R&D engineer.

In the case I discussed at the bank, the young engineer was forced to sell his patent to a large international company, as no bank would help him finance development of his business. Now he's the divisional manager and is responsible for making money from his invention.

Unfortunately many finance directors doubt that they can rely on their R&D engineers for any kind of financial control. However, it is not only possible but it has to happen, as the finance people often have only a vague idea as to what the engineers are up to. The only thing they can do is apply the brake. In some cases such action saves the company, but a race cannot be won without a foot on the accelerator too.

The finance director's role is therefore only to be a coach — not to race him- or herself.

In Chapter 13 we will discuss ways to transfer responsibility for economic control to those who directly manage the business.

# Your comments

*What is the position in your company? Has it delegated responsibility for maintaining financial strength?*

# Balanced growth

## *The way to avoid a STOP/GO strategy*

Expansion which reduces your financial strength will consign you to a stop/go strategy.

You will expand as far as your financial strength will take you and then you are forced to consolidate until you fill the coffers again and can afford a new expansion.

It could therefore be useful to examine more closely the margins needed to provide expansion without reducing your financial strength.

## *The profit margin needed to preserve financial strength*

Let us assume that customers in our business need a credit period of ninety days on average — a very common figure when you are involved in international marketing. Also assume that we turn our stock over twice a year, including work-in-progress — not a brilliant rate but all too common though. Stock is valued at 60% of sales value.

We are not counting on help with finance from our customers in the form of prepayments; this means that about 10% of our turnover can be considered to be 'free loans' (credit from our suppliers, VAT received but not returned etc.).

Let us assume further that we can avoid paying dividends or management fees and that we can build up reserves by making provisions so the tax burden will only be 25% of the operating result after depreciation and finance costs.

Finally, let us assume that we can borrow money at 14%, and that investments in fixed assets represent a depreciation requirement of about 5% of turnover.

What is the lowest profit margin (i.e. result before depreciation and interest, in percentage of turnover) that will retain a financial strength of 30%?

# Room for thought

The lowest sales margin will be about 32%, calculated as follows:

**SALES INCREASE**
*call the increase in sales 100*                                                              100

**INCREASE IN MARKETING ASSETS**
*i.e. the cost of generating 100 of sales*

Accounts receivable (90/360 x 100)=                                              25

Stock ( (100/2) x 0.60) )=                                                              30
*the cost of the stock needed for the 100 sales: this would be 50*
*but it is valued at only 60%*                                                            —

*so the increase in assets is*                                                            55

*Increasing our PnL\* statement by 100 in sales means the*
*asset side of the balance sheet also grows by 55; this means the*
*finance side of the balance sheet must also grow by 55*
*(i.e. 'where does the money to finance the 55 come from?'):*

**TO BE FINANCED BY**
Equity and reserves (30% x 55)=                                                    17
*the bank will generally only lend us 70% (less free loans) so we*
*need to raise this 30% ourselves*

Free loans                                                                                       10

Interest-bearing loans (55-10-17)=                                                28
*the amount we borrow*                                                                   —
                                                                                                  55

**MINIMUM PROFIT MARGIN REQUIRED**
*to raise the 17 needed above*

Increase in equity and reserves                                                    17
*i.e. a net profit of 17*

Add back 25% tax that will be deducted                                         6
*i.e. we will pay 25% tax on the profits*                                             —

Result after depreciation and interest
(17 + 6) =                                                                                       23

Add on interest (14% of 28)                                                           4
*the cost of borrowing the 28*

Add depreciation                                                                            5
                                                                                                  —
Thus:  (23+4+5)=                                                                            32
*Operational result (or gross profit) before depreciation and interest*

---

\*PnL: profit-and-loss.

Few companies can show a profit margin of 32% but, on the other hand, you may not have to give 90 days credit and you may be able to turn the stock over more than twice a year.

But as soon as you export, credit times will become longer and if you need to keep local stock, then the stock turnover rate will be low.

## *Start by cherry-picking*

Why are we so reluctant to charge higher prices for goods and services we export than for those we sell in the home market?

When discussing price sensitivity, the market can be depicted as a triangle. At the top we have the most price-insensitive customers; at the bottom are the most price-sensitive.

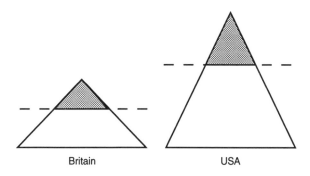

When our company ventures outside Britain, we normally aim for a market share several times smaller than that we hold in our domestic market — at least for a start. Even a small market share in a large market like the USA would probably increase our total turnover substantially. That's why it can be enough to win over just the most price-insensitive customers.

The drawback is that our salespeople will have to put up with relatively more 'no's before they find the price-insensitive customer saying 'yes'. If, on top of this, those salespeople are paid on commission, you can be sure that they will complain bitterly about the unfairness of having to charge higher prices abroad than at home.

## Capital-saving expansion

If the possibility of charging high margins seems unrealistic, the only alternative is to reduce the demand on working capital.

What margin is needed to preserve our financial strength if we can reduce credit time to 60 days and turn the stock over six times per year (with the other parameters the same as before)?

---

**Room for thought**

---

The lowest sales margin will be about 16%, calculated as follows:

| | |
|---|---:|
| **SALES INCREASE** | 100 |
| **INCREASE IN MARKETING ASSETS** | |
| Credit (60/360 x 100)= | 16.7 |
| Stock ( (100/6) x 0.60) )= | <u>10.0</u> |
| | 26.7 |
| **FINANCED BY** | |
| Equity and reserves (30% x 26.7)= | 8.0 |
| Free loans | 10.0 |
| Therefore require (26.7-8.0-10.0)= interest-bearing loans | 8.7 |
| | —— |
| | 26.7 |
| **MINIMUM PROFIT MARGIN REQUIRED** | |
| Increase in equity and reserves | 8.0 |
| Add back 25% tax deducted | <u>2.0</u> |
| Result after depreciation and interest | 10.0 |
| Add on interest (14% on 8.7) | 1.2 |
| Add depreciation | 5.0 |
| Therefore gross margin before depreciation and interest (10.0+1.2+5.0)= | —— |
| | 16.2 |

A margin of around 16% is therefore enough, just over half the margin required in the first example.

## Conditions for balanced growth

There are two ways to achieve balanced growth:

1. High delivery performance and generous credit terms based on achieving high margins; or

2. Lower delivery performance and less generous credit times (perhaps even with the help of both suppliers and customers to contribute to the financing), combined with lower margins.

If you can combine the second alternative with high margins, you get enormous resources for business development.

What you **cannot** afford to do is to have high delivery performance and generous credit times at low margins.

At least, not for very long.

## Your comments

*If you were to increase sales by £1 million (or an appropriate, significant figure for your company), how much would you need to invest in credit and stock, including work-in-progress? Could you make such an expansion without reducing your financial strength?*

# Can the capitalists count on profitability?

*The owner's alternative interest*

We have found that international expansion is capital-intensive. It is natural for owners and investors to hesitate before giving us more money if we cannot show profit on what we already have.

An important tool for controlling business development is therefore the 'alternative' rate of interest.

As we have seen, owners' expectations can vary between 15% and 25%. Variations outside these limits can occur, for two main reasons. First, public companies may be allowed lower profitability, citing public interest. Second, a business may, for strategic reasons, want to 'milk' a mature business unit by requiring very high profitability — in order to be able to reduce the requirements on a young developing unit.

When you try to increase your market share you are often forced to make soft investments, written off in one year, as we have discussed earlier.

When we discuss the owners' profitability requirements it is vitally important to use the right definition of profitability.

*Return on capital employed*

The concept for profitability most suited for controlling business development is 'return on capital employed' (ROCE). ROCE is defined as

$$ROCE = \frac{\text{Total INCOME} - \text{Total COSTS}}{\text{Total ASSETS} - \text{'FREE LOANS'}}$$

Total INCOME should include all financial income, whereas Total COSTS should include all costs inclusive of depreciation but excluding financial costs. If we included these costs, we would have to exclude the loans from the denominator and would get the wrong profitability concept. We would then be discussing profitability on equity.

The importance of using the right profitability concept will be discussed more thoroughly in Chapter 13.

Deducting FREE LOANS from the denominator is a recent phenomenon. You could of course consider adjusting the profitability requirements and disregarding the effects of free loans, but I wouldn't recommend this, not least because free loans can be a strategically important variable.

Such loans not requiring interest often consist of trade creditors, VAT not yet returned, income tax not yet returned etc. These 'loans' often constitute 10% of turnover and, if you can negotiate pre-payments from customers, they can make up more than 20% of turnover.

## *Profitability per capita*

When you choose ROCE as a control variable, it indicates that the bottleneck in business development is capital.

The more the company develops towards being a knowledge or post-industrial company, the more reason there is to complement profitability calculations with another concept: 'profitability per capita'.

In a pure consultancy, you might as well say profitability per consultant. This does not mean that other employees are unimportant but just that capacity is centred around the income-generating individuals.

In other service organisations, you can simply express it as average profitability per employee.

Profitability per capita is defined as:

$$\text{'Profitability'} = \frac{\text{Total INCOME} - \text{Total COSTS}}{\text{No. of 'strategic heads'}}$$

In this case, Total COSTS should include the alternative interest on capital employed, as the capital is no longer the basis for comparison.

Why you should not calculate based on the actual financial costs will be covered in Chapter 13.

The idea behind a yardstick based on 'profitability per head' is that if you do not make enough per consultant, this is an indication not to employ more — as this only will lead to unacceptable levels of risk exposure.

## Invisible profitability

As discussed previously, there are invisible assets in the balance sheet due to the tendency to write off investments in international business development in the year they occur.

If investments in international business development were treated in the same way as investment in machines and equipment, profitability could be improved considerably.

## Long-term profitability

We have argued that a single year's profitability should not determine how we direct our business development, but profitability should be judged over a period that includes a full economic cycle. It should be long enough to spread the variations in development investment.

As analysis should be forward-looking, the profitability we can read from the annual accounts is only a starting point for our forecast and this can give us some very tricky questions:

- Is our bad result this year dependent upon a particularly deep recession or have we lost market share?
- Are we the victims of problems inherent in the economic cycle or because of structural changes in the industry?
- Will the new, half-finished system ever work?

It is very tempting to exaggerate the importance of historical information for such an analysis. If we do, we are guilty of the same lapse of logic as the drunkard when he was found crawling on all fours under a streetlamp:

'My dear man what are you doing here?'

'I'm looking for my key.'

'Are you sure you lost it here?'

'No, but this is the only place I can see well enough to find it!'

The question is not whether we are to withdraw our capital from the company or let it stay where it is — in a post-industrial company (especially) you cannot remove the capital without destroying a large part of the business.

Instead the question is — do we want to risk throwing good money after bad?

---

## Your comments

*What profitability requirement do you have in your company? How is it defined? Is it applicable to individual years or is it seen over a full business cycle?*

# Attack or consolidation?

*Always begin by carrying out a health-check*

We know that our goal should be satisfactory growth but this is subject to two restrictions:

1. Growth should be balanced;

2. Growth should be profitable in the long term.

Before we start the analysis we have to complete the health check below:

| RESTRICTION | TARGET | REALITY |
|---|---|---|
| Profitability | 20% | 12% |
| Financial strength | 30% | 24% |

*The difference between a red and a green light*

The question is whether we have a green light for expansion or not. Should we plan for an aggressive business strategy aimed at taking market share from competitors? Or do we have a red light — meaning we should concentrate on making more money from the market share we have already?

Perhaps we are in such a precarious position that we should retreat by giving up market share. This could be done by milking a segment, through price increases for example.

How would you interpret the figures given in the example above? Would you try to take market share or would you consolidate your position? Or would you retreat?

You cannot answer these questions without knowing something about the business behind the numbers.

The only people capable of answering these questions are technical people and their marketing managers. Only they know if they have invested enough in systems and markets and whether they are now prepared to sit back and let the money roll in. They will know if the expansion will require an explosive growth in marketing assets, or if it can be made self-financing by negotiating long delivery times and pre-payments from customers.

But, of course, they can't do without a good managing director either!

## *Hand on heart*

I have an interesting side job; I am a visiting professor at the Thammasat University in Bangkok.

In the Masters course we run for employed postgraduates, we recently had a marketing director for Coca Cola as a student. During a break she came up to me and said that the example about red and green traffic lights couldn't be used in Bangkok.

'You see,' she began, 'here in Bangkok, green light means GO and red light means GO FASTER!'

Hand on heart, have you always stopped as soon as you've seen the traffic lights change?

My red/green analysis should be viewed in the same way; not to drive as if you are in Bangkok, but to weigh the danger of losing speed against the danger of being caught crossing a red light.

If the reality behind the figures in the example above is that we are ready to harvest with very small additional investments in marketing assets, I would say GO.

If, on the other hand, the systems are only half-finished and we are still not fully established in the markets we expect to serve, and if to do so will require huge additional investments in marketing assets, I would shout STOP. To start an expansionist business plan under such circumstances must be considered 'reckless driving'.

The importance of the red/green analysis can hardly be stressed enough in post-industrial business development. Due to the large proportion of 'soft investments', the financial result will be disastrous if the money will not take us all the way. If we have to retreat before we can harvest, the result will be an enormous waste of capital.

An important realisation in post-industrial business development is that any financial analysis has to be forward-looking, as the question is not whether to take invested capital out of the company, but whether to continue investing.

The profitability that can be expected depends to a large extent upon the estimated future investment needs of our business development.

Interestingly enough, it could even be that a high profitability historically is a result of an earlier decision to harvest, whereas business conditions now may be so bad that there is no future for the company.

The most significant constraint on our expansion is our financial strength or, rather, how the bank perceives our financial strength. If we cannot borrow when we need to, to go all the way with our development, major loss of capital is inevitable.

Applying this directly to a company like OPERATOR, if a business unit manager like Leonard Hewson doesn't get the money to launch his latest system this year or next, it could be too late — the Japanese or other competitors will have beaten him to it.

## Your comments

*Carry out a 'health check' for your company:*

| RESTRICTION | TARGET | REALITY |
|---|---|---|
| Profitability<br>Financial strength | | |

# Summary

Uncontrolled growth is a recipe for disaster as international expansion requires huge amounts of capital.

If we cannot ask our customers to pay in advance, or cannot ship products until we ourselves have bought and produced, we need to invest in credit and stock.

Eventually, we will have to increase our fixed assets, but first we need soft investments in business development, investments that are conventionally written off in one year.

No capitalist worth his or her salt will throw good money after bad, so if we cannot show profitable use of the resources we already have, we will not get more and hence will be unable to expand.

The profitability concept most likely to fit the post-industrial business development is ROCE ('return on capital employed'). In this concept, the result after depreciation — but before financial costs — is seen in relation to total yielding capital. Most investors are satisfied with a return of 20%.

One group of investors, the banks, are assured of a 'dividend' whatever the result, which means that they are happy to lend us money as long as we can show such collateral in the balance sheet as will satisfy them. Provided we have not borrowed excessively we can solve a liquidity crisis by asking the bank for an increased overdraft.

The measurement we use to predict the willingness of the bank to lend us money is the concept of financial strength. Here, we relate our own equity to the total capital used in the company. Most companies consider 30% to be adequate.

If you cannot expand in a way that retains your financial strength, you should make sure that it is well above 30% before attacking your competitors' market shares.

Chapter 4

# Follow the narrow road

*This chapter considers why post-industrial companies should aim to be world leaders in specialised niche markets. It will make little difference whether we have plans to export or not; our international competitors are not likely to let us have an uncontested home market.*

## 'Mind your own business'

*Let's discuss business*

When discussing business development it is important to talk about genuine business and not administration. Let me give you an example.

> On one occasion OPERATOR Ltd.'s Board was happy to see that stock levels were falling and that stock turnover was rising. Unfortunately, however, debtors were on the increase.
>
> 'Well,' a member of the Board asked, 'isn't it better to have money in debtors than in stock?'
>
> 'Yes, at least it is one step closer to cash,' someone commented.
>
> Although this may be true, what had happened in this case was simply that some of the stock had been transferred from the parent company to a subsidiary in the United States.

We need to distinguish between business units and administrative units. We must talk 'business' all the time.


## Which business?

Soon enough, when discussing business development, you have to discuss smaller units than the company as a whole. To go to the extreme of discussing separate products and customers may, on the other hand, prove unfruitful.

The way we split the operation up into business units is crucial to future business development.

For some time, OPERATOR Ltd. had separated their Consultancy Department from Hardware Sales. The reason for this separation was that this way it was easier to charge and be paid for consultancy.

Customers often thought that consultancy should be a part of Hardware Sales:

'No, I will not pay you to find out how many flame controllers should be installed, or where, or how they should be programmed — that's all part of your sales job to sell them to me!'

Keeping Consultancy organisationally separate from Hardware, however, led to a loss of synergy and soon the consultancy people began specifying competitors' products.

But with the general decline in the importance of hardware, whilst the importance of system specification, programming and problem-solving increased, the situation for the Consultancy unit became untenable. As consultancy work was considered to be a service, charged by the hour, it was soon realised that selling complete systems — including hardware — would be far more profitable, so the two units were again merged.

The way in which a company defines its business has a crucial influence on the direction of business development. It is the very starting point; business that hasn't been defined cannot be discussed.

If, as in the case of OPERATOR, the business has been defined as two strategic units, one selling components and the other consultancy, then you cannot discuss sales of systems and problem solutions. (Here I consider a system to be components plus knowledge/consultancy.) We will soon return to the question of how best to define business units.

## Whose business should we be discussing?

Before we continue we should clarify at which level we should be discussing business.

Are we discussing the business of the United Kingdom in general; of The Concern plc; of Subsidiary Ltd; or perhaps of a business unit called, say, Foundry Systems, within Subsidiary Ltd?

There are many and varied ways to structure the hierarchy of a business. I cannot claim to have found the definitive solution but I have found that the following structure makes sense:

-   'The CONCERN' — sometimes referred to as Head Office — is a strategic unit with independent financing.
-   'The COMPANY' is a strategic unit with an independent balance sheet and profit-and-loss account.
-   'The BUSINESS DIVISION' is a strategic unit within a company which can be assigned a reasonably independent balance sheet and profit-and-loss account. It will also control reasonably independent resources to pursue a distinct business idea.
-   'A BUSINESS AREA' is an additional geographical level within a business division.
-   'A BUSINESS UNIT' is a combination of a business division and a business area.

You will notice that the outline becomes more diffuse the further down the hierarchy we go, but we have to live with that. If your mental well-being depends on crystal-clear definitions, you shouldn't be doing business.

However, perhaps some examples from OPERATOR will sharpen the picture.

OPERATOR Ltd. is a company in the Concern called INVESTOR plc.

OPERATOR has an independent balance sheet and profit-and-loss account but does not have independent financing. INVESTOR can, at any time, take the kitty out of Charles Armstrong's hands by asking for 'management fees'— something of which Charles is only too well aware. He gets very upset if someone in Head Office even so much as whispers about taking money away from him — money that he has earned. Unfortunately, that is the whole point of owning more than 90% of a company; namely to be able to direct the money to whatever activity needs it most, irrespective of how the money has been earned.

OPERATOR's subsidiary, OPERATOR Inc. in the USA, is not a company, nor is it a business unit. This subsidiary is purely an administrative unit — it does not have an independent profit-and-loss account or balance sheet. This was also seen in the example, earlier in this chapter, where stock was being moved from the UK to the USA; certain products were stocked in the USA whereas others would be stocked centrally in England and be sent by express carrier to the different subsidiaries.

OPERATOR Ltd. and OPERATOR Inc. are like Siamese twins in other respects as well: there are, for example, common resources for system development and customisation.

A business division is what you might call 'Profit Area Metal', which has operations in both the UK and the USA.

The business division 'Metal' can then be subdivided into business areas such as UK, Europe, USA etc.

'Metal (USA)' is then, for example, a business unit.

As many businesses depend on shared resources, it is important to be able to express a consolidated picture of a business division like Metal. This does not always happen, sometimes with lamentable misunderstandings as a result.

So whose business should we be discussing?

We are discussing **the company's** business.

The ideal situation might be a company consisting of only one business division. That is hardly ever the case, however, as it is very difficult for an organisation to 'stick to its knitting'!

## *Stick to your knitting!*

In the post-industrial economy, with its overwhelming range of business opportunities, it is tempting to get involved in too many different activities. Demands in the market are virtually infinite and we are in a position to satisfy them to a larger and larger degree. As much of what we are selling is knowledge, it may be tempting to try to solve a particular customer's other problems without actually deciding strategically that we should be in this line of work.

Modern technology invites diversification into too many application areas.

I am old enough to have worked with computers programmed with punched cards — with that technology, you were forced to prioritise your activities severely. To this day I can remember the excitement I felt when we were allowed to start working with the new generation of electronically programmed computers.

'I can do whatever you like,' the programmer offered. 'If you're prepared to pay for it,' he added. But I soon forgot that part.

Too easily do we misjudge or forget what it costs to use even a known technology in new application areas. And, once we have dared to take the leap out to one export market, it is all too easy to start working in too many overseas markets.

Another way of stretching your resources is to provide too large a part of the added value yourself. Should we really do everything ourselves, or use subcontractors?

The saying 'stick to your knitting' is more appropriate today than ever before. The trend in modern business development is towards focus upon a narrow specialist or niche market.

The name of the game is to become 'world leader in your own special niche market'.

Remember that this note of caution is appropriate even if you are not planning to work overseas. International competition is coming here!

## Your comments

*Is your 'COMPANY' independently financed, or can it be said to belong to a 'CONCERN' that can request management fees? Do you aim to become a world leader in a niche, or do you try to serve too many markets?*

# Finding your international niche

## *The three dimensions of business development*

In his classic book *'Corporate strategy'*[1] the well-known marketing strategist Professor Igor Ansoff has pioneered a model for business development which, as far as I am concerned, can be represented by the following drawing.

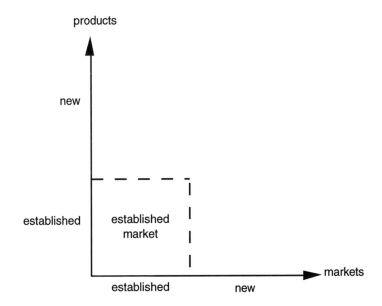

In the post-industrial economy, we should add a third dimension to the model: applications, in line with our earlier discussions.

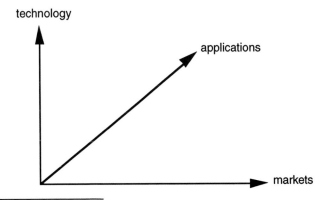

---

[1]Ansoff, I., *'Corporate strategy'*, McGraw-Hill, 1965.

In the case of OPERATOR Ltd., the axes in the diagram might look like this:

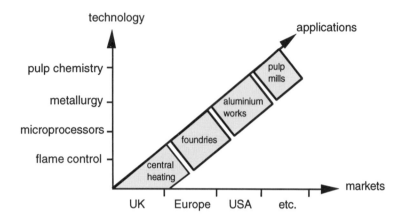

It all started when the founder wanted to make the combustion process in his central heating unit more efficient, which he achieved using a sound knowledge of energy technology and new microprocessors that had recently been introduced.

The solution was very effective and before long all the home-owners in the vicinity wanted a system installed. The founder and his technical friend decided to make energy-saving a full-time occupation and offer flame controllers to all home-owners.

Soon they found that the really big energy savings would come from industry and, with contacts in the foundry business, flame controllers for foundries became the first application.

The leap out on to the international stage happened without much prior consideration. One of their customers wanted systems installed in its subsidiaries in Germany and Sweden. The expansion into North America seems pretty logical but how the company ended up in Singapore, no-one can explain.

After a few years, however, the market's interest in saving energy began to decline. To justify the prices of the new systems, all the advantages of higher quality foundry output had to be used. Soon a good knowledge of metallurgy became more important than a fine grasp of energy-saving technology.

What is now stretching OPERATOR to the limit is the unfortunate — or fortunate — fact that one of the female engineers is married to an engineer working in the paper industry. He was impressed by OPERATOR's systems and suggested they could be used in pulp processing. The paper pulp industry consumes vast amounts of energy to boil off water and the way it is carried out has a great impact on the quality of the pulp.

Ansoff's model offers a very good guide to business development. What do you think of the following logic?

First of all you should have a relatively large share of those market segments where you are already established. Thereafter the question is whether you should:

1. Use present technology and application experience in new markets (export); or

2. Develop new applications with known technology for the existing market (product development); or

3. Improve existing applications for existing markets through new technology (product development); or

4. Enter new markets with new technology applied in new ways (diversification).

## Increasingly heavy invisible investments

The last of the four ways of developing business above, diversification into new markets with new technology, could well be a short-cut to committing commercial suicide, as it leads to multiple large but 'soft' investments, conventionally all written off in the first year.

## Increasingly light deliveries

As the material or hardware content of our deliveries (products) decreases, the value per kilogram increases and local markets, previously protected from international competition, may become threatened.

A product I would like to use as an example is beer. Brewers used to have a market which was free from international competition — it simply wasn't profitable to transport water.

Today, when you buy beer, you pay not just to quench your thirst and perhaps for a slight intoxication, but also for more diffuse feelings of image portrayed in extensive brand advertising. The rugged outdoor life in Australia is used to sell lager, mostly to men. I am certainly not convinced that the success of these brands in the UK and Europe is due to the taste alone!

Naturally you do not necessarily send water from Australia; you just export the methods to brew this particular beer locally and the right to use the brand name, which is the result of many years of soft investments in international consumer advertising. (Only when customers are prepared to pay an even higher price premium do you export genuine Australian water as part of the beer.)

## Increasingly rapid development

When a company like OPERATOR introduces a new system, there will be a window of perhaps three years in which to capitalise on its advantages, before the Japanese or other competitors enter the market with another, even better, system. At that point it is important to have made enough money to be able to develop something new, ready for launch.

## The home market may not be enough

The high demand for development in a post-industrial market, combined with the short life-cycles experienced, means that our home market may not be large enough to ensure a satisfactory return on our investment.

## Go for an international niche strategy!

The conclusion is to concentrate on one technology for one or a few applications and bring this knowledge out to well-chosen international markets. You should aim to be the world champion in a well-defined niche market.

# Your comments

*Try drawing a diagram like that for OPERATOR near the start of this chapter, for your company.*

# Is the EU really so important?

*Choose the right market!*

Once you have decided to start an overseas expansion, it's important to make a rational choice of geographical market.

Too often the choice is dependent on a managing director who loves going to Paris, or that nice Austrian you met on the 'plane.

## *The 'natural' choice*

For companies based in Britain, the most natural choice of market (i.e. the closest) after the UK and Eire will be western Europe; in fact in all probability the northern part of the European Community.

Once that is accomplished, the 'Latin barrier' is the one we need to climb. This is not so much a barrier of language as one of culture and the way to do business. I had a very talented friend who once wrote a book called *'Lion talk'*. He chose the title from a quotation: 'even if we could speak the language of the lions, we wouldn't understand what they were saying'.

Having broached the subject of cultural barriers, you can understand why many companies, having penetrated North America, tend to give Japan (one of the largest markets in the world) a wide berth and rather take on markets such as Australia.

## *The barriers to 'knowledge trade'*

The more knowledge-intensive our products or deliveries are, the less relevant are the old trade barriers of distance and border controls.

## *'Knowledge is easy to carry but hard to explain'*

The more knowledge-intensive our deliveries are, the more difficult they are to explain. This is why language difficulties and cultural barriers should not be trivialised.

Many consider that the purchase of sophisticated technology — like flame controllers — should be entirely rational. Unfortunately, this is far from the truth. In reality, the more complicated a purchase is, the more room there is for personal preferences.

Just suppose that you are the purchasing manager charged with buying oil for your factory's central heating. You have always bought the oil from your brother-in-law, who happens to be the local sales representative for one of the main producers. The oil he sells is a very well-defined product, highly standardised with no distinguishing features.

One day, a slick salesman from a competing oil producer visits your office and gives you a quote that undercuts your brother-in-law by 20%. Even if you can't stand the new man, given that all else is equal, you have no choice but to buy from him — just imagine if your MD found out that you had missed a chance to reduce costs by £10,000 per year.

Yet, if the same unsavoury character came along with a suggestion to install a complex electronic system that might save you more than the system your friendly relative offers, I'm sure you could find a reason to choose your relative; you can trust his service support, he has local references etc.

I do not mean to imply that you would deliberately choose the 'wrong' system, but the more complex the offer is, the more room there is for subjective considerations.

## *Locally-added value is increasing*

The more knowledge-intensive our deliveries become, the more the local proportion of the added value increases, as we have seen before.

A benefit of this is that the more our systems are made locally the less we are dependent on currency fluctuations and affected by customs duties.

## *A 'cultural' choice of markets*

In view of this discussion, I maintain that there is a 'cultural' alternative to the 'natural' choice of markets mentioned earlier. Why not sell knowledge- and information-intensive products and systems where they are easiest to explain?

In such a case we should still penetrate our home market first. After this we might choose the other 'Anglo-Saxon' markets of the USA and Canada. Australia and New Zealand are terribly far away, and as markets are relatively small, but as we have noted before, knowledge is easy to transport. And what about the rest of the Commonwealth? In some countries the cultural similarities are high, in others less so.

After this Anglo-Saxon cultural grouping, the Saxon culture is the easiest for most people to cope with. This includes Germany, Austria, part of Switzerland, Holland, part of Belgium and the countries of Scandinavia, amongst others.

After having penetrated the Saxon markets, we begin to hesitate at the 'Latin barrier'. However, this is not insurmountable, even if an international company arranging conferences recently had to back out from France because a new language law there requires that English conferences held in France must provide simultaneous translation into French — even if the subject was international marketing strategies in this case! When it was held in Germany instead, no interpretation facilities were provided.

When you do finally go to Japan, and other south-east Asian countries, with your business, you may realise that the French and Italians are the brothers and sisters of the British, speaking from a cultural point of view; perhaps the Latin barrier was not so hard to climb after all.

### *'They call me in the middle of the night!'*

There is one dimension that runs contrary to the line of reasoning above — the east/west dimension. The main problem is time differences.

You have to admit that the fact that we in Europe have no office hours in common with Vancouver or Los Angeles makes business discussions somewhat more difficult. The spread of fax machines and especially networks, however, is beginning to ease this problem.

### *'Our situation is unique'*

'Our situation is unique' is a statement that consultants like myself almost always hear when visiting a new client. And, of course, to an extent, it is true; every company is unique.

If your company delivers systems to the forestry industry you probably do not have a big market in the UK.

I cannot decide for you how to attack the world market — you have to decide for yourself. I can only hope that you have found some useful pointers in this chapter.

---

## Your comments

*Do you know on which principles your company chooses its geographical markets?*

# Sticking to your knitting

It is easier said than done to stick to your core business idea. There are many temptations along the way.

One particularly appealing argument to diversify is that of spreading risk.

> 'Yes — but wouldn't it be good not to be wholly dependent on the foundry business?' suggested Carol's husband to Charles Armstrong, trying to convince him of the advantage of developing a system for the pulp industry.
>
> 'We have systems for aluminium foundries too,' said Charles Armstrong, for once slightly hesitant.
>
> 'Ah yes, but that's still metallurgy.'

## *The spreading of risks should be handled 'upstairs'*

The key question is whether it is safer to be weak in three areas or strong in one.

I can understand the argument that it is dangerous to become too dependent on one line of business, particularly one that is very sensitive to fluctuations in the economic climate.

But where in the hierarchy should these considerations be made?

They should definitely **not** be made in a business division and preferably not even at company level, unless the company is completely independent financially.

I think the decision to spread risk (or not) should take place at the concern level.

# Are we ready to split up?

## *To develop but retain a unique competence*

The challenge in post-industrial business development is to develop whilst retaining a unique competence, in order to become the best at something customers worldwide would like to buy and pay well for.

But how can we start up new activities if all established companies refuse to try something they haven't done before? The lone inventor isn't a favourite customer with the banks, as we saw earlier.

## A chance we cannot miss?

Charles Armstrong would like to 'stick to his knitting' but he's under pressure from his business unit manager, Carol Parker.

'This is a chance you have to take!' Carol tells him, visibly agitated. 'Kenneth has both an order from his old company and money, in the form of a Government grant, for the development.'

## A greenhouse for ideas

I think Charles Armstrong would decide to hire Carol's husband Kenneth as manager of a newly formed business division.

If so I cannot fault Charles here; promising new activities must get a chance and it is difficult for them to develop properly in any environment other than within an established company.

Established companies can work as a greenhouse for new ideas and activities. However, the new ideas should have a certain affinity or synergy with the established activities of the company. If not, any overseeing 'corporate body' may have a tendency to reject them.

## Too successful?

The reason for the corporate body's rejection is likely to be the risk that the new activity becomes a 'black hole', absorbing so much in the way of resources and attention that the original business suffers.

The best approach is to define the new activity as a business division as soon as it has left the experimental stage. Ideally, it should have a profit-and-loss account and a balance sheet from the start. This way the resources spent can easily be controlled.

You can hardly expect fantastic results from the new activity straight away, largely because of the soft investments necessary to establish a position in the market.

Really you only have two choices; either you capitalise these investments and require 'normal' profitability levels, or you use the same write-off principle as for the rest of the company and adjust the profitability requirement.

I prefer to adjust the profitability requirement for two reasons. First, this avoids discussion about what should and shouldn't be capitalised in the balance sheet. Second, it is easier to get an overview of total profitability if the same principles are applied across the board.

## Your comments

*Has your company persevered with one business idea or do you have much 'sibling rivalry'?*

# Summary

Post-industrial products weigh very little and have a high value per kilogram. Protected local markets hardly exist any more. Large development investments are written off in one year and short life-cycles can make the home market too small to achieve an acceptable profitability. Therefore, it is important to become 'world leader' at something. The more narrowly we specialise, the easier it is to reach this position.

Few companies have managed to concentrate on a single activity. The reason normally given for this diversification is the desire to have more than one leg to stand on. A more honest explanation might be that it is fun to try something new.

The spreading of risks should be done at the 'Concern' level. By 'Concern' we mean a group of companies with independent financing. By 'Company' we mean an activity with an independent balance sheet. Overseas subsidiaries are rarely companies, as their profit-and-loss accounts and balance sheets are intimately linked to those of their parent company. They should instead be seen as administrative units, partly there to satisfy the host country's legal requirements.

If the company cannot concentrate on one activity, it is important to separate different activities into business divisions capable of being assigned reasonably independent profit-and-loss accounts and balance sheets. These business divisions can then, in turn, be subdivided into geographical business areas.

Finding the optimum strategy for geographical expansion also requires proper study. Too often chance and luck play a role in choosing the geographical market to develop.

Chapter 5

# At the first crossroads

*In this chapter, we will consider an important question:
Can we fully implement market-based pricing?*

*A post-industrial market price is not clearly defined;
unlike the price of four star petrol, it can be what we make it!*

After getting the 'green light' to expand, we've now started the journey on the straight and narrow road to international success. Soon enough, we reach the first crossroads.

## How do we set our prices?

*Pricing — a worthwhile activity*

In the marketing of industrial and other products, pricing is not always given much attention. It is often a matter for the finance department to decide, even though price is the most difficult competitive weapon to handle.

Price is by far the most expensive competitive weapon.

By how much do you need to reduce prices in your business to make your customers notice? Is 5% enough? Calculate how much 5% of your company's turnover represents, then think of what could be done with this amount if you invest it in development, or more salespeople, or better delivery.

Price is the only competitive weapon within the function:

$$\text{REVENUE} = \text{VOLUME} \times \text{PRICE}$$

## *Pricing methods*

In principle, there are two ways of pricing.

One is the cost-based pricing method. The starting point is the cost, to which we add a suitable profit to derive the price we charge the customer.

Traditionally this is the most common method.

The second method finds out what the market is prepared to pay. This method could be called market-based pricing.

Many companies already claim to have market-based pricing. However, I bet that if I could go into the organisation and 'scratch the surface' of the pricing process, I would find that the market orientation is a thin veneer over a pricing that, in all important aspects, is a product of the finance department's cost calculations. I often find that market-based pricing is a euphemism for being prepared to give discounts.

I will admit that there are exceptions to the rule but personally I haven't come across that many.

# Your comments

*How is pricing controlled in your company? How do you decide which business deals are profitable?*

# Passive cost-based pricing

*The trap of cost-based pricing explained*

I started my career teaching management courses by categorically ruling out costs as a base for pricing. Now, years later, I realise that I hadn't considered all aspects of business, because in some cases it **does** work.

My 'ritual murder' of cost-based pricing was based on the following little model:

| cost+cost+cost+profit | | market price |
|:---:|:---:|:---:|
| 100 | | ? |

How could anyone believe that he or she could sit in an office adding cost upon cost finally to arrive at the price the customer wants to pay? The customer doesn't actually have to pay us at all if he/she doesn't want to; customers are not in the least interested in our costs. The only thing customers are interested in is solving their own problems, one of which may be their perception that our price is too high (or possibly even too low).

Customers naturally do not pay us because we have high costs; they pay us for the benefit they believe they will obtain from our offer, compared with what they would have to pay for the same or a similar offer from someone else.

If we make the mistake of starting from the cost side, on the left in the diagram, we can only go wrong. We can either calculate our price to be above the market price or below it.

| cost+cost+cost+profit | | market price |
|:---:|:---:|:---:|
| 100 | | 85 |

Assume the 'market price' is 85. This means the price we have calculated ourselves (100) is above what the customer is prepared to pay. This, however, does not necessarily harm us, as the market will automatically correct our mistake; if we want to sell, we have to reduce our price.

| |
|---|
| cost+cost+cost+profit |
| 100 |

| |
|---|
| market price |
| 115 |

But, instead, now assume that the market price is actually 115, whilst for some reason we think 100 is enough. Perhaps we have been able to replace costly electro-mechanics with inexpensive electronics, for example. Now we have calculated our price to be below the market price.

'Well, is that such a problem?', I used to say, tongue in cheek. 'Won't the friendly purchasing manager tell us that we have quoted far too low, and that we could or should charge more like the other competitors?'

No, of course they won't. Instead they will say 'It's too expensive'. All professional buyers do — it's part of the job description: 'When you hear a price, say it's too expensive!'

To drive this lesson home, I used to tell of a buyer who also happened to be a good friend of mine. When he said his automatic 'it's too expensive', I asked him: 'What do you mean? Every time I give you a price you say it's too expensive. Why?'

'Haven't you understood?', came the reply. 'It's the most profitable activity there is. It hardly takes any time to say, and I normally get at least 5% reduction. You calculate the hourly rate on that!'

At this point, none of my management students doubted that market forces could only correct our pricing mistakes downwards — never up. If we want to get paid for our 'golden eggs', we have to make the change to market-based pricing.

## A sobering thought

I must have performed this ritual murder of cost-based pricing hundreds of times in different groups with participants of varying degrees of commercial and financial understanding. None had ever been able to find a flaw in the logic.

Yet, as time went by, I started to feel slightly uncomfortable.

When I had worked as a consultant for a few years, I was forced to admit that most companies, some of them among the biggest in the country, were using cost-based pricing. I could no longer avoid the question of how they had been able to grow so successfully if they had built their businesses on the wrong pricing method.

How was it that all these skilled managers had missed this one fact that I and few others seemed to have found out?

Naturally there had to be a logical explanation to why so many successful companies started 'at the wrong end'.

The reason for using cost-based pricing could not be to cover their costs. With this logic, you could nip into the accountancy department one dark night and add a few hundred thousand to the company's costs and we would immediately start making more money.

## Why cost-based pricing can be so efficient

I had to read widely and give a lot of thought before I found out when cost-based pricing can be appropriate. Among the books I found helpful were Michael Porter's *'Competitive strategy'*[1] and *'Competitive advantage'*[2].

Among the things I learned was that there are very few generic strategies to choose from. Looking at my own experience, the choice can be narrowed down to two different marketing mixes:

| **'Price competition'** | **'Differentiation'** |
|---|---|
| 1. PRICE | 1. PRODUCT |
| 2. PLACE | 2. PROMOTION |
| 3. PRODUCT | 3. PLACE |
| 4. PROMOTION | 4. PRICE |

[1]Porter, M., *'Competitive strategy'*, Free Press.
[2]Porter, M., *'Competitive advantage'*, Free Press.

## *Price competition*

Provided we sell simple material products, where the value is easy to determine, you cannot avoid price competition.

Simple and material products is possibly the wrong description. Sometimes you cannot avoid price competition on very sophisticated products, and nothing prevents services from being judged primarily on price.

The key question is whether we can differentiate our product (from others) or not, to make the customer realise that our offering is different from those of our competitors.

Take a very sophisticated product like petrol — is there really any significant difference between, say, Shell and BP petrol?

So when it comes to petrol, **price** is the main competitive weapon. It is so decisive that price changes are headline news. Price differences of a few per cent will have significant impact on market shares.

In second position we find **place**. You don't drive to the other side of town in order to save just a few pennies, or at least I hope you don't.

The product is so well defined that whatever you do in the way of promotion is aimed more at reinforcing the brand name than explaining the virtue or value of the product. Think about a slogan like 'Let ESSO put a tiger in your tank'.

As long as you stick to the marketing mix I have described as 'price competition', there is really no reason to abandon cost-based pricing. It is easy to apply and it automatically guides you correctly.

Even if we spend 13% of our turnover on price competition (i.e. if we are satisfied with a price of 100 although the market price is 115), we have invested this potentially large sum on the competitive weapon ranked highest in the marketing mix — namely **price**.

When word gets around that a supplier is delivering top-grade products at 100 instead of 115, we will get more orders than we can handle. As we have full coverage of our costs at 100, we can expand our capacity. With a high relative market share we will climb the experience curve and become even more efficient, perhaps allowing us to reduce our price to 95, which in turn gives us an even higher market share etc.

## *Optimising competitive advantage on price*

What you do when you calculate your prices is to optimise your competitive strength on price.

Once you have set aside what you think you need for the other competitive weapons, you calculate the lowest price at which you can survive.

This is a very rational method provided that price is the most decisive factor in getting the business.

---

### Do you need to read on?

*If your company mainly competes on price and if you have flexible costs and a 'green light' for profitability and financial strength (i.e. to expand profitably), then you really do not need to read on. You can continue as before. But is that really the case?*

---

# Active market-based pricing

## When the old conditions are no longer valid

The marketing mix I call 'price competition' is intimately tied to the old industrial society. The product life-cycles were long enough to make most products and services well-known and well-defined. After a period of time there was not much to focus on other than price.

Examples of the old stable conditions are the electro-mechanical telephone exchange, petrol and the lathe itself.

But then the post-industrial or information society arrived as everything became computerised; not only the electronic telephone exchange but fuel-injected engines and high technology manufacturing tools.

## Who wants to buy a 5-year old computer system?

To illustrate that it is time for a rethink, I could ask the question: 'Who wants to buy a 5-year old model of a computer system at a 20% discount?'

'You can buy my old 086 or 286 machine at half price if you want — I'm sure you can run it for another ten years. But I'm afraid you can't run the latest programs on it and you can't communicate with many other computers.'

## Optimising your competitive strength on development

Who thinks that IBM, or any other computer company, for example, could win the battle for market share were they to sell today's PC systems in three years' time, just knocking 20% off the price?

In the post-industrial society, you have to optimise competitive strength on development. Nobody wants to buy yesterday's products and services even at reduced prices.

That is why the 'price competition' marketing mix does not work well in development-intensive industries, and neither does cost-based pricing. We have to change strategy and go for 'differentiation'.

Instead of investing 13% of our turnover in price competition, we should charge the 115 in the earlier example, not to pay out dividends to our shareholders, but to increase development even further.

If we do have the green light to expand, we have reached the profitability level the shareholders request from us. Now they want satisfactory growth and this can only be achieved by increasing market share.

The traditional way to increase market share has been to offer a lower price, but that marketing mix is no longer valid. We have to win by investing in development, quality and service instead. That is why we have to charge market prices.

## Costing is a science but pricing is an art

After nigh on a hundred years of tradition in fixing prices by calculation, charging market prices is easier said than done.

An American is said to have coined the phrase 'costing is a science but pricing is an art'. He was of course referring to market-based pricing.

One obstacle to overcome is that many companies believe they already apply market-based pricing, so they lack an awareness of the problem.

# The meaning of a post-industrial market price

## The common mistake

'Yes, but that's not the way we do it' is a common reaction when I describe the disadvantages of cost-based pricing. 'We know the market well and set our prices accordingly.'

But do we really know the market price for sophisticated products and services?

## *The market price of cotton yarn*

When we say 'market price', some people think of it as the point in a diagram where supply meets demand.

For the market price to be that well-defined, however, it is necessary to talk about a bulk product, like cotton yarn, for example. There are very few possibilities for differentiation once you have decided on the quality of the cotton. The spread in prices around the average will be minimal — probably only a few per cent.

## *The market price of a multi-operational manufacturing cell*

If we look at a sophisticated technical system such as, for example, a multi-operational manufacturing cell, with associated production control systems, the variation around an average price can be very large — easily 20 per cent.

The value of such a system to the customer varies greatly, and the perceived value is also dependent on our ability to communicate the advantages and profits that our system generates for the customer, in comparison with other competing systems.

The market price is what we make it.

# Do you need to read further?

*If your company's prices are so well-defined that you could look up today's rates and if, in addition, you have flexible costs and a green light on profitability and financial strength, then you hardly need to read on. You can continue as before — but is this really the case?*

# It is in the customer's interest that we charge the right price

## 'We only charge what it costs'

The fact that a price is the result of a calculation does not mean it seems reasonable to the customer, particularly when compared with other alternatives. We have all seen how companies, if protected from competition, let their costs increase as there has always been the opportunity to pass these costs on to the customer. But then perhaps the word 'customer' is the wrong term in these cases!

## The market economy and morality

Many people are keen to say that they have ruthlessly bargained down a price in a negotiation. Few speak of having successfully charged a high price.

I do not want to be misunderstood here. I am totally against a pricing strategy where a customer's naivety or weak position is used as a reason to overcharge.

'Unfair pricing' generally carries the connotation of someone charging too much; rarely does it lead you to think of the poor supplier having been bargained down to an unprofitable price level. The weak position of a supplier even seems to be a good reason to squeeze him or her further.

The reasoning behind how a seller should set his or her price (assuming that a buyer is powerless to influence it) could be said to be one of the fundamental pillars of Communism which, it has to be said, is not really in favour as a mainstream political ideology in the 1990s. The majority of political parties in the western world now embrace the free market economy, where sellers and buyers can freely agree on prices.

One precondition for such a system to work well is that effective competition must exist. Most developed countries have legislation to prevent activities which will reduce competition too far.

In a market economy prices reflect negotiations between buyer and seller in free competition. This is neither moral nor immoral.

If you still want to pursue the question of morality and immorality, you could say that a market-based price, starting from the customer's relative perceived value, is the most 'moral' way to price a product.

## Pricing in the industrial society

Demand in the industrial society was largely focused on material and often standardised products, regularly leading to the conclusion that a low price was the most important competitive weapon. The company with the lowest price tended to get the most customers. Increasing market share meant even bigger economies of scale, which could further reduce costs and therefore further reduce prices, in turn increasing market share, and so on in a cycle.

Thanks to free competition, prices were under constant pressure towards a level where suppliers just about covered their costs, including a market interest (i.e. a reasonable return) on working capital.

## Pricing in the post-industrial society

In the post-industrial or information society, buyers can generally afford to ask for the best rather than the cheapest. The most important competitive weapon therefore becomes **development**.

Whoever introduces the best products and services will win market share, leading to increased income and even more resources to invest in development, provided they charge prices the market is willing to pay.

This is how the market can give the most effective company the opportunity to go on investing in development and growth.

Thanks to free competition, development costs are continually being pushed up to a level where suppliers just about cover their costs including market interest on working capital.

## The paradox

If customers are looking for increasingly sophisticated products and services, rather than low prices, they should accept the argument that 'we have to cover our development costs'.

Unfortunately, we cannot rely on this. The main pre-occupation of buyers is still to solve their own problems, one of the most important of which is reducing their costs. The task of the buyer is to go through the market looking for the supplier who can give maximum benefit at the lowest price.

The paradox is that if the seller cannot resist price pressure, the customer's long-term needs will be neglected because of short-term considerations. If the seller invests his or her resources in price competition, new solutions of the kind the buyer wants to see may never be developed.

## We have to protect ourselves — and the customer — from short-term gain

The conclusion is that it is in the interest of both supplier and customer to charge a price proportional to the increased value we create for the customer through the unique properties of our solution.

As we saw in Chapter 3, most investors will be satisfied with a 20% return on capital employed. What happens if we get prices that will exceed this ROCE figure? We should reinvest the 'surplus profit' in further development and support, which will benefit the customers in the end.

## Your comments

*How do you feel when a customer asks you how much your product is? If you feel a certain discomfort, you have not dealt with your own objections. Don't believe for one moment that the customer cannot read your body language and see that you think it's a rip-off. If you don't think that the customer is being given good value for money, you are trying to charge immoral prices, irrespective of any calculations you may have made*

*Does your company give value for money? Or are the prices you charge so low you cannot give value for money in the future?*

# Summary

Once we've set out on the 'straight and narrow', we'll soon come to a crossroads: 'Shall we choose cost-based or market-based pricing?'

Provided we sell products or services that cannot be differentiated — e.g. petrol — we cannot avoid price competition. The product is so well-known and well-defined that we have no answer to the question 'Why should I pay you more?' A lower price will always pay off in a bigger market share.

Under these circumstances, it is quite correct to maximise competitive strength on PRICE, which is achieved through calculation of prices. After having decided how much is needed to just about cover the necessary investments in PRODUCT, PLACE and PROMOTION, you calculate the lowest price on which you can survive, i.e. give full cost coverage including interest on working capital.

When the company moves into the post-industrial market and tries to become world leader in a niche market, where expertise and good service are more important than a low price, market-based pricing becomes necessary.

The transition to market-based pricing is a major step. Many companies think that they have already taken that step just because they are prepared to negotiate when necessary. True market-based pricing is something completely different, however.

Market-based pricing starts without preconceived ideas as to what the market should be willing to pay. Finding this out is not the easiest thing in the world but it pays to try.

A common misconception when talking about market price is to think of the economist's definition: the point where demand and supply curves intercept. This definition is only true for bulk products such as, for example, cotton yarn. When you sell sophisticated products and services, the market price will be what you make it.

How do you then apply market-based pricing in a post-industrial sense? We shall develop that further in Chapter 9 — 'Price development'. But we are already at another crossroads — 'What market prices should we accept?'

Chapter 6

# At the second crossroads

*As we decided to apply market-based pricing at the first crossroads,
when we reach the second we hardly have any choice at all.
We have to use market-based business strategies.*

*It is no longer obvious that we should follow the old rule that 'every
business deal must carry its own costs' (it might even have to carry
more!). We must become more flexible in judging profitability,
without becoming more inclined to give in to every application of
price pressure.*

We have a green light for financial strength and profitability and we
have gone some way along the narrow road to becoming world leaders
in a niche market. We have passed one crossroads where we decided
to apply market-based pricing. We are now faced with the next: 'which
market prices should we accept for business?'

## Which market prices should we accept?

*Don't confuse market prices with business selection!*

When we applied cost-based pricing, in the olden days, the two
problems:

1. 'What price is the market willing to pay?'; and

2. 'Which business deals are we willing to accept?'

were effectively one and the same.

When you apply cost-based pricing, you optimise your competitive strength by calculating the lowest price you can accept. Any customer prepared to accept this price is welcome to buy from us.

That's how simple it is! You don't need to separate pricing and business selection within the old industrial mass-marketing strategy.

When you apply market-based pricing, you are in a completely different situation. You then start at the other end, first by finding out what prices the market can be persuaded to pay for our solution. But this does not mean you should necessarily do business at these prices. It is necessary to judge the impact on our profitability of each business deal.

## *Full-cost and contribution calculations*

A traditional way to assess business deals is to make product calculations using a full-cost formula. A full-cost calculation includes not only the costs directly attributable to the object of the calculation (be it a product, service, job etc.), such as materials and components, but also the company's indirect overhead costs, such as salaries, buildings, maintenance etc. The distribution of these costs can be calculated in a more or less sophisticated way.

The least sophisticated way is to allocate overheads to the object of our calculation in proportion to its sales value. The most sophisticated method that I know of is 'activity-based costing', in which a thorough analysis of all cost drivers is used as the basis for cost allocations. Either way it is a job for the finance department, not for the Business Engineer. It is enough that he or she understands the difference between a full-cost calculation and a contribution calculation, and roughly how each of them works.

A contribution calculation looks only at the additional income from the business deal, and deducts from this income only those costs that would disappear if the business deal wasn't done. This calculation is often used as a last resort in utilising overcapacity.

The following table summarises which costs are included when calculating profitability using these two methods:

| Full-cost | Contribution |
|---|---|
| Sales overheads | |
| Administrative overheads | |
| Production overheads | |
| Material overheads | |
| Direct labour | |
| Direct material | Direct material |

## *Hybrids*

There also exist hybrids where you deduct more than just the direct material costs from the revenue. These are sometimes indexed; e.g.:

$\text{Contribution}_2 = \text{Revenue} - \text{Direct material cost} - \text{Direct labour cost}$

The higher the index number the more costs have been included; a high index suggests you are approaching a full-cost calculation. There are no fixed rules on how to index such hybrids — they vary from company to company.

My recommendation is to deduct and number the contribution concepts in an order following the relative ease with which you can adjust the costs in an overcapacity situation. The lower the index number, the closer you are to a contribution calculation.

## *Should we use full-cost <u>or</u> contribution calculations?*

In OPERATOR, Charles Armstrong doesn't hesitate.

'The only sound basis for judgement of profitability is a full-cost calculation. When you start disregarding overhead costs you are starting along the road to financial ruin. Whatever the marketeers say, all costs have to be covered and if we don't get a profit on top of that, we won't survive in the long run!'

But even he is known to have made exceptions.

'Yes, but I always live to regret them!'

# 'A desperate situation'

## *Orders were drying up*

OPERATOR was in a difficult position a couple of years ago. It was October and the situation was as follows:

|  | £M |
|---|---|
| CAPACITY COSTS p.a | −10.0 |
| incl. depreciation and financial costs | |
| TOTAL ORDER REVENUE | |
| from Jan. up to and including Oct. | |
| of 'System Gamma' £10.0 M @ 60% CR*= | + 6.0 |
| (contribution from £10.0 M of orders) | |
| (Full cost £9.5 M) | |
| of 'System Alpha' £6.7 M @ 58% CR*= | + 3.9 |
| (Full cost £5.5 M) | |

\* CR stands for Contribution Ratio which means the contribution as a percentage of the sales value

'Contribution' = income − costs of materials and components

Charles Armstrong was very happy with system Alpha which showed a net profit margin ((revenue − full cost)/revenue) of £M (6.7 −5.5)/£M 6.7 = nearly 18%. He was not so happy with the Gamma system, which was only showing a meagre 5% (=(10.0 −9.5)/10.0)net profit margin.

But the real problem was that no more orders were coming in!

Nick Brown, head of production, complained that he needed orders 'yesterday'. Time to fill the capacity for the last two months of the year had almost run out.

'If I can't order components immediately, my people will have to play cards up until Christmas!'

Leonard Hewson was chasing his sales engineers 'with a blowtorch', but nowhere were there any orders to be found.

Charles had made it clear that he did not want speculative production.

If no more orders came in, the result for the year would be a deficit of £0.1M after depreciation and finance costs. This would mean that as Managing Director of OPERATOR Ltd., he would have to show an Annual Report in the red.

## The Italian order

This was the situation when the export manager Neil Johnson came home from Italy (he was fired some years later because Charles thought he made such 'crazy' deals).

'Listen, I've got a major order for a special version of system Beta! The margins are somewhat squeezed but we haven't got much to do right now....'

This is how it would look if we accepted the Italian order for the Beta system:

|  | £M |
|---|---|
| CAPACITY COSTS p.a. | −10.0 |
| incl. depreciation and financial costs | |
| TOTAL ORDERS | |
| from Jan. up to and including Oct. | |
| 'System Gamma' £10.0 M @ 60% CR*= | + 6.0 |
| (Full cost £9.5 M) | |
| 'System Alpha' £6.7 M @ 58% CR*= | + 3.9 |
| (Full cost £5.5 M) | |
| 'System Beta' £1.2 M @ 25% CR*= | + 0.3 |
| (Full cost £1.3 M) | |

# Should the export manager be fired or promoted?

*Would you have accepted the order for Beta?*

## *Is the Italian order a loss or the profit of the year?*

'The margins are somewhat squeezed,' mimicked Charles Armstrong. 'That must be the understatement of the year! It's a loss-making deal —the income doesn't pay the full cost.'

Another way to see it is that the contribution of £300,000 which would result from the Italian order for Beta would change the result from loss to profit.

Many people would have accepted the Beta order.

'Well I believe we have to do business next year too!' Charles Armstrong reacted angrily when this reasoning was brought up. 'What do you think happens to the price levels for Alpha and Gamma when it becomes known to our customers that Beta can do the same job? I'll tell you — the price pressure will be so strong that even if we sell our total capacity next year we won't cover our fixed capacity costs.'

'But Charles, didn't I tell you that the systems will be installed in Italy?'

'Let me tell you something Neil. Has it escaped your attention that our largest customers have a trade association meeting every spring in Amsterdam? Don't you think they'll discuss your Italian order?'

'No, I know they won't. It's a technical meeting and, anyway, even if they do, they can't claim that Beta and Gamma are the same system. You see, I've blocked some of the functions the Italians don't need, so to them Beta is a special system.'

'But how do you know there isn't another order for Alpha or Gamma due in soon?'

'Is there?'

'Well, you never know.'

'But you said yourself that it's '5 to 12' in terms of filling the capacity for the rest of the year and nobody knows of any other orders that are on their way in. Do you really want to miss this chance of keeping us in the black?'

'Sometimes you have to play it cool and not be tempted to do bad business. I think you're far too quick to reduce prices.'

It is not easy here to say who is right.

Perhaps Charles is stuck in an old way of thinking. He should realise that with high fixed capacity costs and no speculative stock build-up, the company becomes more and more like an airline; it has one fixed capacity departure daily which needs to be filled.

On the other hand, perhaps Neil is too ready to reduce prices.

But then Charles has given Neil commission on all export sales 'to get him going'. If he can land this order he will not have to go anywhere for years — his commission is based on the sales value, not the profit. Why should Neil be prepared to take risks by insisting on a ten per cent higher price from the customer?

## *Full-cost and contribution calculations can lead you astray*

Full-cost calculation is an excellent tool for business selection in the old industrial market, but not in the new post-industrial market. It assumes what we want to achieve: a full capacity utilisation at, on average, good prices. If you can count on constant, full capacity utilisation at all times, perhaps producing for stock in order to absorb fluctuations in demand, then the assumption is correct.

In post-industrial business development the full-cost calculation does not serve business selection well, as full capacity utilisation can seldom be taken for granted. A modern technical firm customises its offerings to the customer and therefore cannot produce for stock in speculation.

But full-cost calculations can be used for the following:

1. an indication of average profitability;
2. a basis for judging capacity requirements;  and
3. partially for stock valuation.

Unfortunately, you cannot fully trust the contribution calculation either. It only gives you accurate guidelines if there is no alternative use for the capacity. With effective business development, this should rarely be the case.

## *The right way — using an opportunity marginal calculation*

The contribution calculation can, however, be used as a springboard for an 'opportunity marginal calculation'. How this can be made is best illustrated through examples of market-based business selections; we will look at these in the next chapter.

# Summary

Cost-based pricing is often linked to passive, cost-based business selection. This is normally stated as: 'every business deal should carry its own costs'.

Once the decision has been taken to adopt market-based pricing, normally the selection of business also needs to become market-based.

Knowing what the market is willing to pay does not mean that we have to do business at this price. The question becomes: 'how do we determine which business deals to accept?'

The traditional rule of each business deal carrying its own costs is only true if we have flexible costs or the possibility to create flexibility through stock build-up.

In a situation with a large proportion of fixed capacity costs and a high risk of obsolescence, the company's situation becomes more and more like that of an airline. Profitability will become highly dependent on high capacity utilisation and, therefore, also on flexibility in business selection.

A word of warning — flexibility must never lead us to tying up production capacity with unprofitable business as soon as the market puts a little pressure on price, or with business that will ruin the overall price level.

How to achieve flexibility aimed at high capacity utilisation without speculation in stock and at an acceptable price level is the subject of the next chapter.

# Chapter 7

# Post-industrial business selection

*In this chapter we will develop business selection into a structured art by using a consistent 'opportunity cost' concept. This will lead to a more advanced tool for business selection than simply full-cost and contribution calculations —*
***the business opportunity matrix.***

Consider the product range below, simplified for demonstration purposes. Whether these numbers come from OPERATOR or not is immaterial, but as treatment of them is the central theme of this chapter I have dropped the tinted panels used hitherto when discussing business specific to OPERATOR.

| System | Price | Variable unit cost | Contribution₁ (CB₁) | Contr. Ratio₁ (CR₁) | Sales int. | Contr. Ratio₂ (CR₂) | Full cost | 'Profit' |
|--------|-------|-----|-----|-----|-----|-----|-------|------|
| ALPHA | 12,000 | 5000 | 7000 | 58% | 8% | 50% | 10,000 | 2000 |
| BETA | 8000 | 6000 | 2000 | 25% | 4% | 21% | 8500 | −500 |
| GAMMA | 10,000 | 4000 | 6000 | 60% | 6% | 54% | 9500 | 500 |

**Contribution₁** = Price - Variable unit cost
**Contribution Ratio₁** = Contribution/Price
**Sales interest** = Variable capital cost, see Chapter 3
**CR₂** = Contribution Ratio₁ - Sales interest
**Profit** = Price - Full cost

Alpha is our golden egg. Gamma is normal business at normal prices. Beta is the product you can always use to fill up capacity, assuming that you can accept the low price.

Which system is the most profitable?

---

# Room for thought

---

Perhaps I should not have asked the question so bluntly; it is impossible to answer off-hand in a technology-intensive, post-industrial company.

In the traditional economic environment of a company producing commodities that could be stocked — where capacity utilisation could be taken for granted — then Alpha is clearly the most profitable system, whilst Beta generates a loss.

If we had used passive cost-based pricing in a situation with reasonably flexible costs, our decision would have been fairly straightforward. We would have stopped selling Beta and reduced our capacity. If the capacity adjustment had taken some time to implement, we would have used the temporary overcapacity to build up stock of Alpha and Gamma.

But this is not the situation with which we are faced.

The company in this example, OPERATOR Ltd., markets advanced systems for saving energy; systems that comprise a large proportion of software and service support. The cost structure is fixed as the majority of employees are skilled technical experts. The company has invested large sums in product development and market penetration, which must be considered as sunk costs.

So the question instead has to be whether or not to stop selling the 'unprofitable' system Beta.

What would your recommendation be?

## Room for thought

Again the question is not easy to answer without better knowledge of the situation at hand.

Suppose, for example, that Beta is a necessary subsystem for Alpha; then, if we refuse to sell Beta (or price it excessively high) we risk losing sales of Alpha.

However, in this case Beta is not a subsystem of Alpha (the last paragraph was just a 'what if...' to illustrate different aspects of business selection). Customers do often choose us not for a single product though, but because we can present a total solution to their problems. If you make one part of your offering unattractive you may find the customers reject the system as a whole.

This, however, is not the case for OPERATOR. Alpha and Beta are sold independently.

But what if sales of Alpha and Gamma just about cover our capacity costs, resulting in a break-even situation? Can't the incremental contribution from Beta then be regarded as profit — not as a loss?

Charles Armstrong considers such incremental discussions to be dangerous.

The danger in this example is not in the reasoning, but in the fact that our costs have developed to be increasingly fixed and our income more uncertain. This increasing uncertainty cannot be reduced by spending more time calculating things.

Some people use calculations like a drunk uses a lamp post — more for support than enlightenment.

Instead the new situation has to be met with greater flexibility in the business approach. We must constantly judge the maximum price we can achieve, given the market situation and whether orders could be just around the corner.

Market-based business selection requires a good knowledge of the market as well as of our capacity utilisation and cost structure — combined with a good knowledge of 'Business Economics'.

If you don't know what you're doing, from an economic point of view, when applying the new strategy, it could of course become dangerous. But that's the way it has always been — you have to know what you're doing.

## Clutching at straws

No wonder in our example Charles Armstrong clings to passive cost-based pricing with cost-based business selection, like a drowning man clutches a lifebuoy.

But the less concrete the products become (services are the extreme) and the more rigid is our cost structure, the less safety is offered by the lifebuoy — it becomes more like a straw we are clutching.

In the post-industrial economy 'best possible capacity utilisation at best possible price levels, based on customer-perceived value' has to be our guideline when selecting business opportunities (as opposed to passive, cost-based prices and passive utilisation of capacity), even if such a strategy makes greater demands on both the board and the staff.

## A lifeboat is available: the **business opportunity matrix**

Questions such as: 'Which product is the most profitable?' are not relevant if we apply a new strategy based on market-oriented pricing and best possible capacity utilisation. The question in our example should instead be: 'Should we stop selling Beta?'

Before we try to answer that question we first have to clarify our position. Do we have overcapacity or undercapacity?

A capacity precisely in line with demand assumes that we can tell exactly how the market will grow and where the strength is in our marketing mix. Additionally, we have to be able to read the minds of our competitors and anticipate their every move.

In other words, the normal situation is that our capacity is either too large or too small; we have either over- or undercapacity.

Once we have analysed our capacity situation we have to make our minds up as to whether we are discussing short-term or long-term decisions.

And, finally, we have to know whether we have a red or a green light for investments aimed at gaining additional market share.

The '**Business Opportunity Matrix**' is based on these consider-ations. Refer back to the end of Chapter 3 for an introduction to red and green lights. The crux is that a red light means we need to go for profit in order to strengthen the balance sheet. A green light means the balance sheet is sufficiently strong for us to attack and try to increase market share.

|  | Overcapacity | | Undercapacity | |
|---|---|---|---|---|
| Short term | 1 | | 2 | |
|  | Green | Red | Green | Red |
| Long term | 6 | 5 | 4 | 3 |

Let us see how we can use this decision matrix to find the answer to the question:

**'Should we really accept business on Beta at these rotten prices?'**

The answer will be different depending on which sector of the matrix we find ourselves in.

We will now return to OPERATOR and walk around the six situations in the matrix.

# Short-term decisions where there is overcapacity

*What is 'short-term' and 'overcapacity'?*

When you run a complex operation with skilled personnel you do not hire and fire from one month to another. A short-term decision often means 'within a year' in modern companies.

Overcapacity means we cannot fill our capacity with Alpha and Gamma business at acceptable prices, not that people are sitting idle. If we are not experiencing decision paralysis we can always fill the capacity with 'incremental' business, similar to the Beta systems discussed earlier.

| | Overcapacity | | Undercapacity | |
|---|---|---|---|---|
| Short term | 1 | | 2 | |
| | Green | Red | Green | Red |
| Long term | 6 | 5 | 4 | 3 |

## *Back to that Italian order*

Our task is to come to a decision for the near term in a situation of overcapacity. Would you turn down the order for system Beta that salesman Neil Johnson brought home from Italy?

## **Room for thought**

| | Overcapacity | | Undercapacity | |
|---|---|---|---|---|
| Short term | 1 | | 2 | |
| | Green | Red | Green | Red |
| Long term | 6 | 5 | 4 | 3 |

Most marketeers are quick to accept Beta business in this situation, vaguely remembering a rule from cost/income analysis which said that during overcapacity the rule is to accept business provided it covers more than the variable cost. If, in addition, they have commission on turnover like Neil Johnson in the previous chapter, this increases the tendency to accept business at low prices.

This is to do business without really trying.

This is also the reason why market-oriented pricing and marginal costing is sometimes wrongly seen as an excuse to do low-margin business. It is also a reason why marketing and sales people have acquired a reputation for being irresponsible economically.

## An alternative is to try harder

We have to make an extra effort to sell more of the Alpha and Gamma systems.

Making an extra effort to sell more means increasing marketing activity. This generally costs money, unless people have been sitting around all day. We have to accept that we will have to make extra investments in a useful competitive weapon. Extra marketing activities can be expensive, and when the going gets tough, money gets scarce.

## Competitive weapons available in the short term

The only competitive weapons that can work in the near term are price and promotion (at least within the limits of my creativity).

When Charles Armstrong put the marketing department under pressure to avoid having to accept Neil Johnson's 'Outrageous Italian Proposition', Leonard Hewson (Marketing Manager) suggested an extra advertising campaign to introduce system Gamma to a new market sector (the cost for this was of course ex-budget). He presented the calculation in the next section, using the figures in the table at the start of this chapter.

## *An extra advertising campaign ex-budget*

### AN EXTRA CAMPAIGN FOR 'GAMMA'

<u>Sales target</u>
75 systems x £10,000 =                          £  750,000

<u>Contribution:</u>
75 systems x £6000 =                             £  450,000

Campaign cost:                                   £  –50,000

**'Profit':**                                    **£  400,000**

'Hey, my order is for 150 Beta systems! You wouldn't fill the capacity with 75 systems!' objected Neil, seeing his once-in-a-lifetime record commission disappearing.

'As a matter of fact, it does,' said Nick Brown (Technical Director). 'Gamma takes twice the capacity per system compared with Beta.'

'How do you know?'

'Gut feeling. I am responsible for production, if you remember.'

'You don't have to have a gut feeling to know this,' said Malcolm, the Finance Director. 'You can look it up in the full-cost calculation. The difference between the full cost and the variable cost in Gamma's case is 5500 and in Beta's case 2500 (see the summary table on p. 119). This difference is the part of our capacity costs attributed to the system in question and it is an excellent indicator of how much capacity each system needs.'

'You've misunderstood something, Neil,' said Charles Armstrong gravely. 'Our ultimate goal is not to fill capacity at any price, but to make a profit! But I can't approve Leonard's proposition anyway, because it is based on his so-called contribution calculation.'

| | Overcapacity | | Undercapacity | |
|---|---|---|---|---|
| Short term | 1 | | 2 | |
| | Green | Red | Green | Red |
| Long term | 6 | 5 | 4 | 3 |

Charles Armstrong angrily corrected Leonard's calculation as follows:

**AN EXTRA CAMPAIGN FOR 'GAMMA'**

Sales target:

75 systems x £10,000 =                £ 750,000

*Full cost*  ~~Contribution~~:                    − £712,500

75 systems x £6000 = *9500*           £~~450,000~~

Campaign cost:                        £ -50,000

*Loss!*  ~~'Profit'~~:                        £~~250,000~~

                                     − £12,500

'When in the blue blazes will you marketeers learn that contribution is only half the calculation! The fact is that we have people in production who have to be paid, and I assume that you gentlemen have no intention of giving up your salaries. How many times do I have to tell you that we have to include all our costs before we take the business?'

| | Overcapacity | | Undercapacity | |
|---|---|---|---|---|
| Short term | 1 | | 2 | |
| | Green | Red | Green | Red |
| Long term | 6 | 5 | 4 | 3 |

# Room for thought

*Who is right here?*

| | Overcapacity | | Undercapacity | |
|---|---|---|---|---|
| Short term | 1 | | 2 | |
| | Green | Red | Green | Red |
| Long term | 6 | 5 | 4 | 3 |

In fact both are wrong!

Leonard Hewson's method of using the contribution to defend the costs of the campaign assumes that there is no other way to fill the overcapacity apart from increased advertising.

Charles Armstrong's method of requiring full cost coverage is wishful thinking as it assumes what we are trying to achieve; that is to say full capacity utilisation at acceptable prices.

| | Overcapacity | | Undercapacity | |
|---|---|---|---|---|
| Short term | 1 | | 2 | |
| | Green | Red | Green | Red |
| Long term | 6 | 5 | 4 | 3 |

## *The correct way to calculate*

Here is the correct way to calculate, using an opportunity calculation.

In opportunity calculations we compare different alternatives and choose the alternative that pays our capacity best.

### 1. THE 'GAMMA' ADVERTISING ALTERNATIVE

**ALTERNATIVE INCOME (= Sales target):**

| | |
|---|---|
| 75 systems x £10,000 = | + £ 750,000 |

**VARIABLE UNIT COST**

| | |
|---|---|
| 75 systems x £4000 = | − £ 300,000 |

**CONTRIBUTION$_1$ =** £ 450,000

**Sales interest:**

| | |
|---|---|
| 6% on £750,000 = | − £ 45,000 |

**CONTRIBUTION$_2$ =** £ 405,000

**Campaign cost:** − £ 50,000

**CONTRIBUTION$_3$=** £ 355,000

### 2. THE ITALIAN 'BETA' ALTERNATIVE

**ALTERNATIVE INCOME (= the Italian Proposition):**

| | |
|---|---|
| 150 systems x £8000 = | + £ 1,200,000 |

**VARIABLE UNIT COST**

| | |
|---|---|
| 150 systems x £6000 = | −£ 900,000 |

**CONTRIBUTION$_1$=** £ 300,000

**Sales interest:**

| | |
|---|---|
| 4% on £1,200,000 = | −£ 48,000 |

**CONTRIBUTION$_2$=** £ 252,000

**Campaign cost:** −£ 0

**CONTRIBUTION$_3$ =** £ 252,000

### 1-2 = THE OPPORTUNITY MARGINAL CONTRIBUTION

The OPPORTUNITY MARGINAL CONTRIBUTION =

$$= \text{CONTRIBUTION}_{3\text{Gamma}} - \text{CONTRIBUTION}_{3\text{Beta}} =$$

| | | |
|---|---|---|
| = £ 355,000 | − £ 252,000 | = £103,000 |

| | Overcapacity | | Undercapacity | |
|---|---|---|---|---|
| Short term | 1 | | 2 | |
| | Green | Red | Green | Red |
| Long term | 6 | 5 | 4 | 3 |

## A quicker route to the same conclusion

You could just as well start comparing CONTRIBUTION$_2$ for the two alternatives:

| System | Sales | CR$_2$ | CONTRIB.$_2$ |
|---|---|---|---|
| Gamma | £750,000 | 54% | £405,000 |
| – Beta | £1,200,000 | 21% | –£252,000 |
| | | | +£153,000 |

Thus the difference in CONTRIBUTION$_2$ exceeds the extra cost for advertising Gamma by £103,000 — this is the OPPORTUNITY MARGINAL CONTRIBUTION in favour of Gamma.

## What does the **opportunity marginal contribution** tell us?

This opportunity calculation tells us that an extra campaign to sell more of system Gamma gives a result £103,000 better than selling system Beta at the proposed low price — **assuming that the advertising campaign results in the desired turnover increase** (which Charles Armstrong doubts!).

The two alternatives:

1. filling up capacity with Beta business with a low contribution per system; and

2. extra advertising for Gamma

have different levels of risk.

## Risk calculation

We are 99% certain that we can get the Beta business if we accept the low price, but there are no guarantees that the advertising campaign will succeed. The calculation is therefore not complete without a break-even analysis — the most widely used method to evaluate uncertainty in business life.

| | | Overcapacity | | Undercapacity | |
|---|---|---|---|---|---|
| Short term | | 1 | | 2 | |
| | | Green | Red | Green | Red |
| Long term | | 6 | 5 | 4 | 3 |

The simple question is:

How many Gamma systems do we have to sell as a result of the advertising campaign to cover the cost of advertising (£50,000) **plus the alternative CONTRIBUTION$_2$ for Beta (£252,000)**?

To answer this we divide this sum by the CONTRIBUTION$_2$ per system for Gamma:

$$\text{BREAK-EVEN} = \frac{(£50,000+£252,000)}{(£10,000 \times 54\%)} = 56 \text{ systems}$$

$$\text{SAFETY MARGIN} = 75 - 56 = 19 \text{ systems } (34\%)$$

So if the campaign results in the sale of 56 Gamma systems the advertising alternative becomes as profitable as the low-price Beta alternative. The goal of the campaign is 75 systems, giving a safety margin of 19 systems or 34%, i.e. we can miss the sales target for Gamma by one third.

Is a safety margin of 34% enough?


## Income is lower but costs are up

Charles Armstrong gave a definite thumbs down to investing fifty thousand pounds in an advertising campaign with such a small safety margin.

'I've seen them miss by more than a half before!'

Leonard Hewson became exasperated and accused Charles of business myopia.

'We have only calculated the initial effect,' began Leonard. 'The systems we sell with the help of this campaign will become reference sites, which will help sell other systems without too much effort. And you should consider the psychological importance of avoiding export sales at half the margin you require from my sales engineers.'

'And you don't build a market position with an old system like Beta. You do it with Gamma and Alpha,' supported Carol.

|            | Overcapacity |     | Undercapacity |     |
|------------|:-----:|:---:|:-----:|:---:|
| Short term | 1     |     | 2     |     |
|            | Green | Red | Green | Red |
| Long term  | 6     | 5   | 4     | 3   |

But Charles Armstrong did not change his mind and I can understand him. If the campaign fails, he has to explain to the board how he has managed to miss the sales goal whilst, at the same time, he has also exceeded the cost budget. The board expects him to reduce costs when the income is going down.

Reluctantly, Charles Armstrong had to admit that the Italian order would guarantee that he could face the shareholders with an Annual Report showing a profit, albeit a small one.

| | Overcapacity | | Undercapacity | |
|---|---|---|---|---|
| Short term | 1 | | 2 | |
| | Green | Red | Green | Red |
| Long term | 6 | 5 | 4 | 3 |

# Room for thought

*What would you have done?*

| | Overcapacity | | Undercapacity | |
|---|---|---|---|---|
| Short term | 1 | | 2 | |
| | Green | Red | Green | Red |
| Long term | 6 | 5 | 4 | 3 |

I would not have hesitated, provided I had the freedom to act without the risk of being kicked out by a suspicious board.

I would have developed the business on Gamma — there is no real future in Beta.

## The first reminder of the importance of financial strength

Investing in an extra campaign for Gamma is an investment for the future. Low-margin Beta business is just 'holding the fort'.

This is the way you can argue if you are strong financially and have good liquidity. If the liquidity is strained and the bank hesitates to increase your loans you cannot make these investments for the future. Instead you may be forced to revert to desperate measures such as selling at knock-down prices.

This is the first reminder of the importance of being strong financially; to dare to take risks and do good business even in a recession.

## Have we got the alternatives and their consequences right?

Opportunity costing is relatively simple from a mathematical point of view but very difficult in practice. You have to know the alternatives and their consequences.

In our example, are there only two alternatives to wasting capacity; low-margin Beta business and the extra advertising campaign for Gamma? I am sure there are more, but these were the best the management team at OPERATOR could think of. (We will consider more alternatives later.)

Let us consider the consequences of the Gamma and Beta alternatives in a little more detail.

| | Overcapacity | | Undercapacity | |
|---|---|---|---|---|
| Short term | 1 | | 2 | |
| | Green | Red | Green | Red |
| Long term | 6 | 5 | 4 | 3 |

When we calculated the safety margin:

$$\text{BREAK-EVEN} = \frac{(£50,000+£252,000)}{(£10,000 \times 54\%)} = 56 \text{ systems}$$

$$\text{SAFETY MARGIN} = 75 - 56 = 19 \text{ systems } (34\%)$$

we assumed — quite correctly — that if we miss sales of Gamma we cannot automatically substitute Italian Beta business, as we had to say no to the Italian proposition in order to safeguard capacity for the flow of orders resulting from the Gamma campaign.

But to sweeten the Italian proposition, Neil Johnson had promised Charles that he would negotiate a flexible delivery time, which he duly did. Therefore we do have some flexibility in delaying production for the Italian order for up to six months to make use of subsequent slack periods in our capacity.

## Room for thought

*Does this alter the safety margin?*

| | Overcapacity | | Undercapacity | |
|---|---|---|---|---|
| Short term | 1 | | 2 | |
| | Green | Red | Green | Red |
| Long term | 6 | 5 | 4 | 3 |

A flexible delivery time on Beta certainly has an effect on the safety margin for the Gamma campaign. So now we are not actually risking the entire contribution from the Italian Beta proposition. The only money at risk is the cost of the advertising campaign.

Therefore we should rephrase our question to be: 'How many Gamma systems do we need to sell to pay back the advertising campaign?'

We find the answer by dividing the cost of the campaign by the opportunity marginal contribution$_2$ of substituting two Beta systems by one Gamma system.

Why two Beta systems for one Gamma system?

Perhaps you will remember from our earlier description of OPERATOR that the capacity needed to produce one Gamma system is twice that needed to build one Beta.

So the new break-even calculation will be

$$\text{BREAK-EVEN} = \frac{£50,000}{\{£5400 - (2 \times £1680)\}} = 25 \text{ systems}$$

$$\text{SAFETY MARGIN} = 75 - 25 = 50 \text{ systems } (67\%)$$

This is a dramatic change! Thanks to the flexible delivery times to Italy, Charles can now say yes to both propositions!

## Why not reduce the price of Alpha?

'If we are to compete on price, why not reduce the price of Alpha?' asked Leonard. 'My people tell me we are 10% too high. Given the right price we could sell twice as much.'

'They always say we're 10% too high.'

Yes, that may be true. Dripping water will wear a hole in a stone. The buyers they meet will always tell them they're too expensive, and they are on commission.

But why not give it a try? Let's start by calculating in the traditional way and then correct ourselves.

|  | Overcapacity | | Undercapacity | |
|---|---|---|---|---|
| Short term | 1 | | 2 | |
|  | Green | Red | Green | Red |
| Long term | 6 | 5 | 4 | 3 |

If we let

| | | |
|---|---|---|
| the **present** volume | = | $V_1$ |
| the **present** CONTRIBUTION$_2$ | = | $C_1$ |
| the **new** volume | = | $V_2$ |
| the **new** CONTRIBUTION$_2$ | = | $C_2$ |

we can write

$$V_1 \times C_1 = V_2 \times C_2$$

as a condition for the total contribution before the price change to be the same as the total contribution after the price change.

The equation can be rearranged to read

$$V_2/V_1 = C_1/C_2$$

which means that the new volume at which the total contribution is the same as the old total contribution, relates to the old volume as the old contribution per unit relates to the new.

The equation can once again be rearranged to read

$$V_2 = V_1 \times (C_1/C_2)$$

Using the figures from the table at the start of this chapter, in thousands, $C_1$ for Alpha is 6.0 and $C_2$ is 4.8, as the price reduction would be 10% of 12. (We assume that the price reduction hits the contribution fully, i.e. there are no reductions in commissions etc.)

So if we now insert the contributions (in thousands, to make the equations shorter) before and after the price change into this equation and set the volume to 100, we have

$$V_2 = 100 \times (6.0/4.8) = 125$$

That is to say, a 25% volume increase is needed to compensate a 10% price reduction.

The promise of doubling the volume would therefore only have to be realised partially in order to generate a larger total contribution after the price reduction.

| | Overcapacity | | Undercapacity | |
|---|---|---|---|---|
| Short term | 1 | | 2 | |
| | Green | Red | Green | Red |
| Long term | 6 | 5 | 4 | 3 |

## Again, have we got the alternatives right?

The calculation above is only correct if we can assume that there are no alternatives to increasing sales through a price reduction. However, this is not the case. We have an alternative use for our capacity, namely filling it with Beta systems.

If we take the Beta alternative into consideration, $C_1$ becomes the alternative contribution, at the present price level, of selling an additional Alpha system against selling two Beta systems (Alpha requires twice the capacity of Beta). That is: $6.0 - (2 \times 1.68) = 2.64$.

$C_2$ is then the alternative contribution after the price reduction, which is: $4.8 - (2 \times 1.68) = 1.44$. We now have

$$V_2 = 100 \times (2.64/1.44) = 183$$

which means that we need almost to double sales to reach break-even.

## A second reminder of the importance of financial strength

If, contrary to good business practice in a post-industrial company, price was the preferred competitive weapon, it is important to know that it will reduce the margin available to finance marketing assets.

Price competition requires good financial strength and liquidity.

## Training — better than low-margin business

As a company develops towards a knowledge-driven organisation, it becomes increasingly important to raise the level of competence of the employees. Contrary to what normally happens, we should therefore use a recession, or a dip in demand, to increase training and education activities, as the alternative cost for the employees' time is now low. Unfortunately, the normal reaction in times of falling demand is to reduce training activities, as part of belt-tightening measures, and to some extent I can understand why.

| | Overcapacity | | Undercapacity | |
|---|---|---|---|---|
| Short term | 1 | | 2 | |
| | Green | Red | Green | Red |
| Long term | 6 | 5 | 4 | 3 |

In order to increase training when orders fail to materialise, you need to have:

1. a long-term view of profitability; and

2. distinct training projects planned.

That seems a lot to ask, which is why very few companies see training as a pro-active way of using their employees' capacity in periods of low activity. Another reason is that it impacts negatively on the visible income statement, and thus also the balance sheet, as we do not regard human resources as assets in traditional accounting.

## A third reminder of the importance of financial strength

You need money to be able to run training programmes as an alternative to low-margin business. Therefore you need good financial strength.

## Development is better than low-margin business

Development is the most important reason for success in post-industrial business. Why not concentrate on development instead of doing bad business in times of low capacity utilisation? Development does not just mean product development, incidentally.

Development work on new systems may even be put aside because R&D engineers have to help to sell the systems they have previously developed. Development engineers are perfect extra salespeople. They have a natural authority when selling our new systems as they themselves have developed them.

There are also many other things that need developing; for example, production methods, administration etc.

As we will discuss in more detail in the next chapter, it is important to avoid bottlenecks in any company. Using times of low capacity utilisation to become familiar with each others' work is therefore always a profitable investment.

| | Overcapacity | | Undercapacity | |
|---|---|---|---|---|
| Short term | 1 | | 2 | |
| | Green | Red | Green | Red |
| Long term | 6 | 5 | 4 | 3 |

## A fourth reminder of the importance of financial strength

Naturally we need good financial strength to be able to concentrate on development instead rather than doing low-margin business.

## When we have no other choice

Finally, when all other sensible alternatives have been tried, we may have to accept low-margin business on the basis that 'it's better to get something than nothing at all'.

However, sometimes low-margin business cannot always be found when you need it. It is therefore prudent to keep in contact with this part of the market even when the going is good. There are companies that continue to subcontract part of their capacity even when their own marketing department is clamouring for shorter delivery times.

If you do 'Beta business' it is important to carry it out in such a way that it does not damage the price structure for your high-margin business. This is an important subject, so we will come back to it in later chapters.

## Rigid costs have to be met with flexible business selection

Skill in handling temporary overcapacity is crucial to profitability in a post-industrial company.

Once you have decided upon market-based pricing, the spread around the average profitability becomes significant. This is inevitable and should be seen as a sign that market opportunities have been fully utilised.

An airline once summed up its strategy on market-based pricing in combination with flexible business selection as: 'We have to charge enough when we can so that we can discount when we have to'.

Low-margin business is not very desirable, but it can be worth having to avoid losing capacity permanently, but before we accept it we should do all we can to avoid it.

| | Overcapacity | | Undercapacity | |
|---|---|---|---|---|
| Short term | 1 | | 2 | |
| | Green | Red | Green | Red |
| Long term | 6 | 5 | 4 | 3 |

If our financial strength allows, we should spend additional money ex-budget on market activities for our high-margin systems.

Alternatively, again if our financial strength allows, we can use our capacity to pre-produce orders with long delivery times. This strategy can be used even by pure service organisations, assuming we do not routinely promise shortest possible delivery times.

Speculating by building stock without orders to back it up is rarely advisable due to the high risk of obsolescence in modern business.

Good alternatives to low-margin business are investments in training and development, but these require good financial strength.

When all the possibilities to avoid low-margin business have been exhausted, we have to realise that it may be better to get something rather than nothing.

In summary, the more rigid our cost structure, the more flexibility is needed in our business selection.

## How do we compare a thousand alternatives?

In the real world, comparison of alternatives can be an overwhelming task. We are unlikely to be just three systems and opportunities in the market can be very hard to gauge. Under these circumstances it is often difficult to establish the alternative value of the capacity correctly.

However, in competitive business, it is more important to do the right thing adequately than to do the wrong thing accurately. A rough estimate of the alternative value is often good enough in order to come to the right conclusion when choosing between business opportunities.

## Not every business deal can be above average

The first Tuesday in every month Charles Armstrong reads the profitability statistics, where all the prices in all the business deals are compared with full-cost calculations. When the price is lower than its full cost it is shown in red print, with the responsible person's name highlighted; that person will have some explaining to do.

| | Overcapacity | | Undercapacity | |
|---|---|---|---|---|
| Short term | 1 | | 2 | |
| | Green | Red | Green | Red |
| Long term | 6 | 5 | 4 | 3 |

Is it wrong to hate bad business?

I would suggest not. Exceedingly low prices always require explanation. The problem is that what constitutes bad business depends on the circumstances, as we have seen in the case of the Italian order for Beta.

A significant variance around the average profitability is a sign of health in a post-industrial company. If there was no variance around the average profitability, this would mean two things:

1. At times we are leaving money on the table; and

2. We are missing opportunities to utilise our capacity.

## The 'low quartile'

In a post-industrial company applying market-based pricing and business opportunity selection, the distribution of profitability around the average could look something like this:

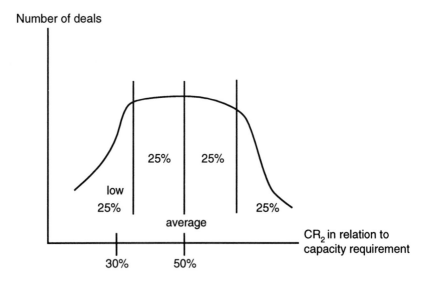

144

| | Overcapacity | | Undercapacity | |
|---|---|---|---|---|
| Short term | 1 | | 2 | |
| | Green | Red | Green | Red |
| Long term | 6 | 5 | 4 | 3 |

This means that we can always assume that there will be a low-end alternative, which we can use as a minimum alternative value of our capacity. The lower the demand, the lower will be the contribution ratio in the 'low quartile'.

Let us return to the question of whether we should spend extra money on advertising Gamma. Let us assume that Gamma is only one of many systems, so that we cannot distinguish one single alternative with which to compare it. The best thing we can then do is to reason as follows:

The $CR_2$ for Gamma is 54%. A random sample shows that the 'low quartile' shows a comparable $CR_2$ of only 30%. The opportunity marginal contribution is therefore $54 - 30 = 24\%$. The cost of the campaign is £50,000. The minimum sales of Gamma must then reach £50,000/0.24 = £208,333 or 21 systems to break even.

The 'low quartile' concept is a very practical and flexible tool. Normally you have a gut feeling for how low the alternative $CR_2$ in the low quartile is. In good times it will not be too low, whereas in bad times it can be close to zero.

If you don't want to rely on gut feeling you can build a statistical system or a sampling system. However, as I have stated, being just about right in opportunity costing is better than being wrong accurately using full costing in a post-industrial company, where capacity utilisation cannot be taken for granted.

| | Overcapacity | | Undercapacity | |
|---|---|---|---|---|
| **Short term** | 1 | | 2 | |
| | Green | Red | Green | Red |
| **Long term** | 6 | 5 | 4 | 3 |

# Short-term decisions where there is undercapacity

*'We should never have taken that low-margin business!'*

'If I get hold of that Johnson, I'll strangle him,' growled Charles Armstrong at the weekly meeting. 'We should never have taken that Italian low-margin business. Where is he?'

'He's travelling,' answered Carol Parker. 'Where did you think he'd be?'

The situation has changed dramatically for OPERATOR. Contrary to forecast, orders began pouring in just after the new year.

'I told you I had several large orders ready to be signed just as soon as the production managers got their new budgets approved. But you didn't want to believe me,' said Leonard.

'This is the last time I listen to Neil Johnson's siren songs about 'at least we get a contribution from the business'.'

Charles Armstrong had conveniently forgotten that he himself had taken the decision to accept the Italian order for Beta systems, once they had been allowed flexible times.

## The bottleneck

'In our situation, we should be very pleased for the Beta order from Neil,' Malcolm (the Finance Director) said, surprising the rest of the meeting. 'Beta needs very little capacity in the bottleneck that has developed in our systems department. This has resulted in ridiculously long lead times on most other products.'

| | Overcapacity | | Undercapacity | |
|---|---|---|---|---|
| Short term | 1 | | 2 | |
| | Green | Red | Green | Red |
| Long term | 6 | 5 | 4 | 3 |

He put a table up on the overhead projector:

| NARROW SECTOR: | Department S | | |
|---|---|---|---|
| | PERSON DAYS | CONTRIBUTION$_2$ | |
| SYSTEM | per system | per system | per day |
| Alpha | 5 | 6000 | 1200 |
| Beta | 1 | 1680 | 1680 |
| Gamma | 4 | 5400 | 1350 |

Note:
The Contribution$_2$ figures used in these calculations are derived from the table at the beginning of the chapter; the person days are the technical director's figures.

## The profitability paradox

'We have a very unfortunate imbalance in our capacity; we have overcapacity in all departments except Department S, where the work is piling up. To add to the frustration, the time for our most profitable product to pass through Department S is so long that Beta, despite its low contribution, is the most profitable alternative.'

## Can we afford (not) to invest to get rid of the bottleneck?

'Yes,' began Carol Parker. 'That's exactly what I pointed out six months ago. We should hire more systems engineers.'

'We've tried, but we only get CVs in from beginners,' Charles replied sadly. 'It takes at least three years for a new recruit to become productive and throughout that time he'll decrease productivity as our more experienced system engineers have to act as teachers and coaches.'

'Why should it be a *he*?' Carol retorted quickly. 'Hire a woman instead — that'll speed things up.'

Charles Armstrong knew the routine well enough not to respond to this. 'Can't we buy capacity from someone else?' Leonard queried, but added immediately: 'But I suppose that would effectively tell our competitors how to beat us. Couldn't we just buy some routine programming?'

| | Overcapacity | | Undercapacity | |
|---|---|---|---|---|
| Short term | 1 | | 2 | |
| | Green | Red | Green | Red |
| Long term | 6 | 5 | 4 | 3 |

'Department S doesn't handle *routine* programming,' said Carol.

'We could more than double the capacity in Department S if we bought the new computer system with robot stations offered to us by CADMIAC,' offered Nick Brown, OPERATOR's technical manager.

'And what would that cost?' asked Charles.

'About a million, but we would get fifteen complete workstations and the whole assembly line robotised. It would also mean that we could...'

'No,' Charles interrupted. 'Don't even think about it.'

But he did think about it.

## What do the board and the bank have to say?

'What do you think the preliminary result we're supposed to show the board on Thursday will look like?' asked Charles.

'Well, it won't look pretty,' said Malcolm, the Finance Director. 'Profitability will be somewhere around 10%, which just about pays for our finance costs. If we break even we should be happy.'

'What about our financial strength, our equity ratio?'

'We might reach 20%.' The targets had been 20% profitability and more than 20% equity ratio.

'Then there's just no way I can propose an investment of £1 million,' concluded Charles Armstrong, shaking his head.

## Another reminder of the importance of financial strength

Again we are reminded that financial strength is crucial for business development. Even if, unexpectedly, the board would agree to an investment of £1 million, they would hardly want to fund it by increasing share capital. A no-profit/no-loss result is not the right basis for issuing new shares.

Nor is a no-profit/no-loss result a good starting point when asking the bank for another loan, particularly if we had already borrowed 80% of our capital needs earlier.

We are forced to conclude that we have a red light and undercapacity. Now we need to draw the right conclusions from this sombre fact.

| | Overcapacity | | Undercapacity | |
|---|---|---|---|---|
| Short term | 1 | | 2 | |
| | Green | Red | Green | Red |
| Long term | 6 | 5 | 4 | 3 |

# Long-term decisions with undercapacity and a red light

*Customers are queueing up but we aren't making any money*

'It never rains but it pours,' Charles complained, in reference to the sad fact that the systems department couldn't deliver what the customers wanted to buy.

'But is it really true then that we can't make enough money this year?' Carol asked.

'It's quite easy to calculate,' answered Malcolm. 'If we're lucky, we can fill our capacity with Beta systems...'

'If we're *lucky*?' yelled Charles.

'Yes — as Beta gives the best contribution per day in the systems department, we'll get the best total contribution if we just sell Beta...'

'But that's crazy!' interrupted Charles, angrily, once more.

'... but even if we manage to do this, the total contribution will only just about cover our capacity costs, plus depreciation and finance costs,' continued Malcolm, ignoring Charles's outburst. 'If, in addition, we have to sell Alpha and Gamma the contribution will be even lower.'

Charles Armstrong did not want to accept this line of reasoning at first but, after some calculations on a flip chart, Malcolm managed to convince him.

'What should we do then?' asked Charles despairingly.

'We have to increase the capacity in the bottleneck without spending the full million. If we can't do that, then we'll have to reduce capacity in other areas to redress the balance.'

'What do you mean by that?' Carol asked nervously.

'Well, there's no point having as many sales engineers if they have nothing to sell, and we wouldn't need quite as many administrative staff either, which means even I would have to lose valuable co-workers,' Malcolm replied sadly.

| | Overcapacity | | Undercapacity | |
|---|---|---|---|---|
| **Short term** | 1 | | 2 | |
| | Green | Red | Green | Red |
| **Long term** | 6 | 5 | 4 | 3 |

## *No money for investment*

'There's no way we'll get any money for investment,' Charles Armstrong told the next meeting. 'I sounded the board out at our recent meeting, but there was no support for my ideas.'

'So what do we do?' asked Nick Brown.

'That depends on you,' Charles answered. 'If you and your people can't reduce the times for Alpha and Gamma in Department S we may have to start laying people off.'

## *Between a rock and a hard place!*

Nick and his people got together for two weeks with the sole aim of solving the problem they had been given by Charles Armstrong.

How they solved it will be revealed in Chapter 12.

## *Why not raise prices?*

Even if productivity is increased considerably in Department S, the capacity does not seem to suffice, at least not within the lead times considered reasonable by customers.

'If we have to turn orders away, at least we can make sure that those we do accept have maximum profitability,' said Malcolm at one of the weekly meetings. 'Why not increase prices for Alpha and Gamma?'

'What, you want us to raise prices too when we've only recently reduced the level of customisation?' Leonard asked, looking at Malcolm as if he was singling him out to make life difficult.

'You won't get orders from all your quotes because of our capacity shortage. You yourself told me that the customers won't accept long delivery times.'

Leonard Hewson had no answer. Not much later Charles Armstrong made sure that he really did raise the prices for Alpha and Gamma.

| | Overcapacity | | Undercapacity | |
|---|---|---|---|---|
| Short term | 1 | | 2 | |
| | Green | Red | Green | Red |
| Long term | 6 | 5 | 4 | 3 |

# Long-term decisions with undercapacity and a green light

*Money enough for investments*

'I'm sure we'll have enough money for investment now,' said Charles Armstrong, beaming when he saw the preliminary figures a year later.

'20% profitability and 28% equity ratio!' Malcolm added eagerly.

'It's not enough,' said Charles, getting a grip on himself. 'I promised myself never again to get too deeply into debt with the banks. We have to wait until we reach 40% equity ratio. Our financial strength is a buffer and it's there to protect us from the hard blows. It should therefore vary with time; if 30% is the average, we should have more than that before we really 'go for it'.'

'Unless we can count on margins that make growth possible without reducing our financial strength,' Malcolm objected.

'Show me those margins first,' Leonard muttered, knowing from experience how difficult that would be.

Not until a year later did Charles Armstrong feel he had 'enough in his coffers' to invest.

The thought of investing in new workstations for the systems department was again under discussion and robotisation of the assembly line was needed more than ever. In the meantime, a few new systems engineers had been hired, but the systems department was still the bottleneck, making lead times 'impossible'.

There were now some fifty system engineers working overtime but, even so, they still didn't have enough hours in the day. And, to their great disappointment, they had failed to keep delivery promises to many of the customers.

If it hadn't been for the systems department, it would have been possible to increase sales by 25% without recruiting or investing in fixed assets.

| | Overcapacity | | Undercapacity | |
|---|---|---|---|---|
| **Short term** | 1 | | 2 | |
| | Green | Red | Green | Red |
| **Long term** | 6 | 5 | 4 | 3 |

Advertisements in the appointments sections of the papers had not produced any results. System designers of this type are very attractive in the market and evidently had not been attracted by the kind of package that OPERATOR could offer.

Charles Armstrong now had to convince the board of the sense in investing £1 million in a new CAD-CAM system and automating the assembly line. He did this using a simple payback calculation:

$$\text{PAY-BACK TIME} = \frac{\text{INVESTMENT}}{\text{INCOME SURPLUS}}$$

## How to 'count the baby away with the bath water'

Filling in the 'investment' figure in our payback calculation is not difficult, but the 'income surplus' is more difficult to predict.

True to form, Charles Armstrong first wanted to see the income surplus as the difference between income and full cost for the additional sales. The way full-cost calculation works in OPERATOR — working in full depreciation and profitability requirements, safety margins etc. — only gives a margin of 5% to justify the investment. This means that the new investment must result in extra sales of £20 million. This, of course, is unrealistic, particularly considering the consequent extra investments necessary in marketing assets.

As usual we have to consider the choices available, rather than spending too much time on a full-cost calculation.

| | Overcapacity | | Undercapacity | |
|---|---|---|---|---|
| Short term | 1 | | 2 | |
| | Green | Red | Green | Red |
| Long term | 6 | 5 | 4 | 3 |

# Room for thought

*What alternatives can you find? How would you calculate if you were Managing Director of OPERATOR (in principle only, not actual numbers)?*

|  | Overcapacity | | Undercapacity | |
|---|---|---|---|---|
| Short term | 1 | | 2 | |
|  | Green | Red | Green | Red |
| Long term | 6 | 5 | 4 | 3 |

I would argue in the following way if I ran OPERATOR. We can justify the investment in two ways, either through reduced costs or increased income. We could possibly add a third; increased safety or security.

The proposed investment would double the productivity in the systems department. As the department's salary costs are £2.5 million per annum, the investment would be paid off in less than a year provided that we either double the turnover or reduce the number of staff.

Laying people off in the present situation is not a realistic alternative. We therefore have to justify our investments by an increase in sales.

We know we could sell more if we had greater capacity in the systems department, but how much more is difficult to say. According to the marketing department, we are losing every second order because of excessive lead times. But this has to be seen in perspective. The excuse of having lost an order because we were 'too expensive' is no longer particularly well received by Charles Armstrong; 'You mean you didn't succeed in communicating the customer value', would be the instant response — another excuse is therefore needed.

Doubling the turnover is unrealistic, but let us assume that we can increase it by at least 25%.

| | Overcapacity | | Undercapacity | |
|---|---|---|---|---|
| Short term | 1 | | 2 | |
| | Green | Red | Green | Red |
| Long term | 6 | 5 | 4 | 3 |

# Room for thought

*What income surplus can we expect if we increase turnover by 25% ?*

| | Overcapacity | | Undercapacity | |
|---|---|---|---|---|
| Short term | 1 | | 2 | |
| | Green | Red | Green | Red |
| Long term | 6 | 5 | 4 | 3 |

The average contribution in OPERATOR is some 50% of turnover. An increase in turnover of 25% means raising income by about £ 8 million and therefore increasing the contribution by £ 4 million.

We would not need to calculate on keeping more than a quarter of the additional contribution as our incremental contribution to be able to claim that we have paid our investment back in one year.

If the above justification for the investment is not sufficient on its own, we would have to turn to 'soft' arguments for support.

| | Overcapacity | | Undercapacity | |
|---|---|---|---|---|
| Short term | 1 | | 2 | |
| | Green | Red | Green | Red |
| Long term | 6 | 5 | 4 | 3 |

# Room for thought

*Which other aspects might you consider for an investment of this kind?*

|  | Overcapacity | | Undercapacity | |
|---|---|---|---|---|
| Short term | 1 | | 2 | |
| | Green | Red | Green | Red |
| Long term | 6 | 5 | 4 | 3 |

You could always consider safety and security aspects.

We are talking about our capacity for system development; something at the very heart of our activity. We have been told that we have a 'heart condition'. The systems department is overloaded and the people are in a bad mood. It can be psychologically damaging if people regularly cannot cope with their workload.

The fact that we are not meeting our delivery commitments can also cost us an arm and a leg.

One reason we couldn't employ the system engineers we wanted was the ageing CAD-CAM equipment in use — these hot-shot engineers want modern toys. This can sometimes mean more to them than their salary.

The systems department plays an important role in general system development and competence build-up. In the present situation, however, they are hard pushed to think more than a week ahead.

Viewed in this way, it is as pointless to calculate the justification for the investment as it is to contemplate whether you can afford an operation to correct an acute heart condition. We have to make the investment if we are to survive.

How do you think Charles Armstrong calculated once he realised full-cost calculations were inappropriate?

| | Overcapacity | | Undercapacity | |
|---|---|---|---|---|
| Short term | 1 | | 2 | |
| | Green | Red | Green | Red |
| Long term | 6 | 5 | 4 | 3 |

# Room for thought

| | Overcapacity | | Undercapacity | |
|---|---|---|---|---|
| Short term | 1 | | 2 | |
| | Green | Red | Green | Red |
| Long term | 6 | 5 | 4 | 3 |

Charles Armstrong's thinking went something like this.

So far the Concern had never turned down a proposal from a strong MD, who could show results, when he wanted to make an investment with a payback time of three years or less.

As the investment in question could be regarded as a computer system, Charles thought it would be best to promise to pay the investment back in two years. He therefore simply said that productivity increases in Department S availed by the proposed new CAD-CAM system and robotisation would be worth £0.5 million per annum. This meant he could present the following investment calculation to the board:

$$\text{PAY-BACK TIME} = \frac{\text{INVESTMENT £1 M}}{\text{ANNUAL INCOME SURPLUS £0.5 M}} = 2 \text{ YEARS}$$

Nobody asked how he had arrived at the income surplus but, on the other hand, there were lots of interested questions about the CAD-CAM system and the robotisation programme.

## Are we increasing profitability or decreasing unprofitability?

This is an important issue which has relevance to many areas, not just the foregoing investment discussion.

Too often you see investments justified by cost reductions of the type: 'Thanks to this investment we can reduce our staff by five people, which means the investment will be paid back in a year'.

Even if it is true that you can dismiss workers, you must also ask whether the operation as a whole is profitable. Instead of increasing profit, we could just be reducing a loss!

We do not have to ask these questions in this specific instance, as we know that we have a green light.

| | Overcapacity | | Undercapacity | |
|---|---|---|---|---|
| Short term | 1 | | 2 | |
| | Green | Red | Green | Red |
| Long term | 6 | 5 | 4 | 3 |

## *A bottleneck around the corner?*

One of the assumptions in our discussion of investment in a new CAD-CAM system for OPERATOR has been that we can increase the sales by a quarter once we have alleviated the bottleneck in Department S.

Is this really true?

Bottlenecks have a tendency to pop up in the most unexpected places. One could very well be waiting around the next corner, preventing us from increasing sales by more than a few per cent, before we are faced with new recruitment or investment proposals.

In OPERATOR's case neither Charles Armstrong nor Technical Director Nick Brown believes that this is likely to happen. Nick has even gone so far as to carry out an analysis of the company's resources and simulate what will happen when the sales go up.

## *Overcapacity in machines but undercapacity in people*

In post-industrial companies, the tendency is to tolerate overcapacity in machines but not in people.

Our most important resource is our employees. Making a person wait for a machine is wasteful, much more so than vice versa, unless we operate in a very capital-intensive industry.

## *Investing in flexibility*

Utilising an overcapacity in fixed assets for increased flexibility requires the workers to be skilled at several tasks. Investment in training is therefore just as important as investment in fixed assets.

| | Overcapacity | | Undercapacity | |
|---|---|---|---|---|
| Short term | 1 | | 2 | |
| | Green | Red | Green | Red |
| Long term | 6 | 5 | 4 | 3 |

# Long-term decisions with overcapacity and a red light

Low utilisation of capacity quickly shows up in the profitability of a post-industrial company with high fixed costs.

If a downturn is unexpected — which it normally is — we can find ourselves in the middle of a costly market offensive with big investments in market and system development. As we have seen, these investments are generally written off in one year, which will hit our profitability hard. In this situation a break-even result is virtually a victory.

If financial strength is good and the company has a well-informed board with cool heads, drastic measures can be avoided. However, this is unfortunately not the case with OPERATOR Ltd.

Charles Armstrong has had bad luck. He has not managed to fill his coffers as he had hoped. The Concern's 'bailiff', as Charles calls his parent company's finance director, has decided to tax him heavily by asking for management fees. Charles can't criticise him for it — other companies within the group have had a rough time so letting OPERATOR keep all its profits could mean an extra tax burden for the Concern.

The board members have not held their positions for that long because of recent turbulence in the Concern. No sooner has Charles 'trained' and created good relations with a member of the board than he or she is replaced.

A unanimous board is now demanding rapid improvement in profitability.

## The temptation

There is a temptation to solve a crisis by overcapacity. Many will claim that all that is needed is to increase sales by a few per cent and this will save the bottom line, since about half the sales income can be seen as an addition to our profit.

| | Overcapacity | | Undercapacity | |
|---|---|---|---|---|
| Short term | 1 | | 2 | |
| | Green | Red | Green | Red |
| Long term | 6 | 5 | 4 | 3 |

If you are at the end of a long recession and have just invested heavily in market and system development, it is easy to give in to this temptation, as it is realistic to expect increasing sales without substantial further actions in the market. However, one condition must be made, and that is that expansion must only take place without losing financial strength.

## Time to gag overoptimistic marketeers

If you are at the beginning of a recession and have no reason to believe that you can harvest, it is time to 'gag the overoptimistic marketeers', as someone once said.

Trying to increase sales under such circumstances just makes matters worse. Additional investments in market and system development are needed; they will be written off in a year and, at the same time, an expansion will increase the working capital.

## A comforting thought

Fortunately, in a post-industrial or knowledge company you do not need to use price as your competitive weapon.

As the market is not overly price-sensitive, a price increase should not necessarily cause a dramatic drop in sales, even if the commissioned salespeople would like to have us believe so.

## Increase prices!

Even if we lose sales as a result of a price increase, we now have another piece of the arithmetic of pricing working for us.

| | Overcapacity | | Undercapacity | |
|---|---|---|---|---|
| Short term | 1 | | 2 | |
| | Green | Red | Green | Red |
| Long term | 6 | 5 | 4 | 3 |

This is OPERATOR's short-term cost structure:

Revenue      100

| |
|---|
| Contribution   60 |
| Alternative cost= <br> Variable <br> unit costs     40 |

In the short term, a post-industrial company is very sensitive to loss of volume as it has a high proportion of fixed capacity costs. If we increase our prices by 10% we can only afford to lose 14% of our volume, as can be seen from the calculation (introduced earlier in this chapter) below:

$$V_2 = 100 \times (C_1/C_2) = 100 \times (60/70) = 86$$

In the long term, if there is the will to change the organisation itself, most of the capacity costs can be influenced. If the alternative cost of working capital (the 'sales interest') is taken into account, adjustable costs are likely to be at least 80% of the income.

Revenue      100

| |
|---|
| Marginal <br> Contribution    20 |
| Alternative cost = <br> Capacity <br> costs we can   40 <br> influence <br> incl. sales <br> interest <br> + <br> Variable <br> unit costs     40 |

| | Overcapacity | | Undercapacity | |
|---|---|---|---|---|
| Short term | 1 | | 2 | |
| | Green | Red | Green | Red |
| Long term | 6 | 5 | 4 | 3 |

If we calculate with 80% of the total income as being adjustable costs, after a price increase of 10% we afford to lose a third of our volume and still have the same result as before the price increase. See the calculation below:

$$V_2 = 100 \times (C1/C_2) = 100 \times (20/30) = 67$$

## Shrink the organisation!

If you are lucky, increasing prices and refraining from hiring new people and borrowing may be sufficient. Remember that price sensitivity is often greatly exaggerated.

However, if forecasts show a further decrease in sales this may be too risky and the organisation will have to be shrunk.

In a modern and complex company, a reduction in capacity often means redefining the structure. It is common practice to strengthen the management team with consultants in this situation. To carry through the restructuring, someone with a knowledge of 'turnaround management' can be helpful, with the additional benefit that an external person also brings a certain new perspective to the company's problems.

Even if this competence is available inside the company, workloads are normally too great and day-to-day operations too time-consuming to make full use of this competence without incurring other losses.

## An orderly retreat and a renewed attack

The way uncomfortable decisions, such as the dismissal of personnel, are communicated is extremely important. If the mood becomes one of doom and gloom, the very employees you want to retain may try to find jobs elsewhere.

The measures taken should be seen, and communicated, not as the beginning of the end but as a structured and orderly retreat in preparation for a renewed attack.

| | Overcapacity | | Undercapacity | |
|---|---|---|---|---|
| **Short term** | 1 | | 2 | |
| | Green | Red | Green | Red |
| **Long term** | 6 | 5 | 4 | 3 |

## An aggressive retreat

Let us eavesdrop on OPERATOR's management team whilst they work out a tactic that Charles Armstrong likes to call an 'aggressive retreat'.

'Let's not get depressed! We are basically a very sound company in a strong position and with excellent market possibilities. But at the moment we are sailing in difficult waters and it is time to get ready to meet the storm. We shall have to bring some sails down. In the Territorial Army, we'd say it is time for an aggressive retreat.'

'What does that mean?' asked Carol anxiously.

'We will decrease spending on the most expensive and least effective competitive weapon — low prices — and *increase* spending on systems development and communication of our value to the customer. To play it safe we shall have to plan for a lower volume. This is my master plan.'

| | Forecast | Plan |
|---|---|---|
| **AGGRESSIVE RETREAT** | | |
| | 1997 | 1998 |
| PRICE INDEX | 100 | 115 |
| VOLUME INDEX | 100 | 75 |
| SALES | 75.0 | 64.7 |
| | | (0.75x1.15x75.0) |
| VARIABLE UNIT COSTS | 27.0 | 20.3 |
| | | (0.75x27.0) |
| CAPACITY COST | 44.0 | 35.2 |
| | | (44.0-(0.25x0.8x44.0)) |
| CAPACITY COST INCREASE | — | 3.5 |
| OPERATING RESULT before depreciation and financial cost | 4.0 | 5.7 |
| DEPRECIATION | – 2.1 | – 2.1 |
| FINANCIAL COST | – 2.1 | – 1.4? |
| NET RESULT | – 0.2 | + 2.2 |

| | Overcapacity | | Undercapacity | |
|---|---|---|---|---|
| Short term | 1 | | 2 | |
| | Green | Red | Green | Red |
| Long term | 6 | 5 | 4 | 3 |

'The logic behind this tactic is that we create a higher degree of freedom for action by planning for a lower volume, to which I have assumed that we can adjust our capacity costs by 80%.'

'Why not a hundred per cent?' asked Nick Brown, not recognising this low ambition on Charles Armstrong's part when cost is discussed.

'You can only increase and decrease capacity in fixed increments. For instance, we've got to have a technical director and I assume that you are not willing to cut your salary.'

Nick Brown considered this below the belt so he retorted: 'The same goes for the Managing Director, it is my turn to presume.'

'Maybe it is time to question that. Perhaps it would be a good idea for us all to discuss higher flexibility in salaries in return for higher security.'

'Why have you put a question mark behind the financial result?' Malcolm Stone wanted to know.

'At a lower sales volume, we don't have to tie up so much money in accounts receivable and inventory, unless we increase credit time and/or decrease the turnover rate. Really I have been very conservative here, because when we question our price level with various customers, we should also question credit and stocking at the same time. With money freed from our market assets we can pay off loans and therefore we'll lower our financial costs.'

'I agree fully, so why the question mark?'

'I was a little uncertain about our loans and to what extent we will actually be able to decrease our market assets, but I don't mind changing it to an exclamation mark.'

'But what if the loss in volume will be larger than 25% because of the higher prices?' asked Leonard Hewson, the Marketing Manager.

'Let's check the safety margin.'

'That's easy,' said Malcolm Stone, the Finance Director. 'The increase in the net result is £ 2.4 million. If you divide it by the new contribution ratio, which will be about 70% as the old one was 65%, you'll see that you can lose 2.4/0.7 = 3.4 M further and still break even.'

| | Overcapacity | | Undercapacity | |
|---|---|---|---|---|
| Short term | 1 | | 2 | |
| | Green | Red | Green | Red |
| Long term | 6 | 5 | 4 | 3 |

'Whether you'll lose £ 10.3 or £ 13.4 million as a consequence of an extreme high price policy is very hard to say. We could very well lose more, and then we'll be really deeply in the red!'

'How can you call a 15% price increase in conjunction with a planned sales reduction an extreme high price policy? You might not have to increase a single price!' said Charles Armstrong angrily.

'How can I raise the price level without increasing the prices?'

'Because you say goodbye to the customers demanding the lowest prices, longest credits and quickest delivery times. You can increase the price level by shifting the structure without changing a single price.'

'But wouldn't it be silly to turn away a customer without trying to get a better price from them?'

'Now you're talking!'

'Who says we will lose volume at all, if we only decrease spending on the least effective competitive weapon — price — but increase spending on the priority competitive weapons of system development and communication?' supported Carol.

'Indeed, as I said, my calculation is very conservative,' beamed Charles. 'Let's feel even better by calculating what the result would be if we only lose 15% volume.'

'That's just as easy,' responded Malcolm. 'If we get £ 7.5 million more in sales at 70% contribution our net result will go from minus £ 0.2 to plus £ 7.5 million!'

'But how would we handle that extra volume, if we have cut down our capacity?' objected Leonard.

'That's a point. We might lose a million in increased capacity cost. However, I am quite sure that we can squeeze in quite a lot with the right kind of orders, can't we Nick?'

Nick nodded his agreement.

| | Overcapacity | | Undercapacity | |
|---|---|---|---|---|
| Short term | 1 | | 2 | |
| | Green | Red | Green | Red |
| Long term | 6 | 5 | 4 | 3 |

# Long-term decisions with overcapacity and a green light

Charles Armstrong took the opportunity to cut deep into the organisation; far too deep, many thought.

Charles was of the opinion that recessions are designed to allow companies to go through what he would call 'a bout of fasting'. He considered that good times generate too much flabby fat and cause them to make poor investments, both in people and development.

The sales dip following the price increase proved to be less drastic than expected. In addition, the slimmed-down organisation could handle a much larger turnover than expected.

All of this soon showed up in the profit-and-loss statement. The year following the reduction, the company received a bright green light. There was also some overcapacity which was partly filled with business on the low-priced Beta system.

## The ideal position for attack

A green light coupled with overcapacity is the ideal position for an aggressive marketeer. Now is the time to strike, in order to take market share from competitors.

Not only are our earnings satisfactory and do we have enough money, but we can also expand without heavy investment in personnel and equipment.

## Yet more temptations

This is when the temptation to diversify can become very strong. There are a thousand and one interesting activities we can try when we have money. The temptation to run in all directions at once is great and can lead to disaster.

You may find it useful to read Chapter 4 about the narrow road again to reinforce the message that the right way is to become 'world leader in your niche market'.

And you may also want to ponder the section on balanced growth in Chapter 3 before you rush off.

## *Let your knitting be 'world leader in niche market'*

If we should 'stick to our knitting', what precisely should our knitting be? Simply to continue taking market share in an established market isn't half as stimulating as opening up new markets. It is prudent, therefore, to be sceptical about new suggestions of branching out before we have reached our target share of the markets in which we already compete.

The disadvantage of small shares in many markets is obvious. The slightest hiccough can severely influence the service. As we are small, we will not get spontaneous requests.

Just as it is fun to enter new markets, our engineers like to try new and exciting technologies. Naturally we should be able to afford personal development but if we spend half our time as 'apprentices' we are unlikely to do good business.

At the risk of being accused of repeating myself, I suggest that your knitting be the goal of becoming a world leader in a niche market.

# Do we need full-cost calculations at all?

## *Full-cost calculation is still needed*

Even though I have stated that the full-cost calculation is not always valuable in post-industrial business, I do not believe that this method should be thrown out of the window. As an indication of average profitability, it is a good yardstick on which to base our flexible business selection.

The problem with full-cost calculations can be that they are often misunderstood and miscalculated. The distributed costs are frequently misunderstood to be variable and the more sophisticated the calculation methods — and the less the person doing them knows about the business — the easier it is to miscalculate.

## 'Sanity check' the calculations

There is a simple method to check whether a full-cost calculation makes sense. The method is best explained by an example. Variable costs are mainly bought-in materials, components and services (i.e. those costs that would not be incurred if the budgeted sales were not made), whilst the capacity costs are the remaining, fixed overhead costs.

One year OPERATOR Ltd. had the following budget:

| | |
|---|---|
| budgeted total sales | £60M |
| variable unit costs | –£20M |
| capacity costs | –£34M |
| budgeted result before depreciation | £ 6M |

(what OPERATOR calls **gross profit**)

As soon as the budget was finalised, Leonard Hewson (the Marketing Manager) calculated what he would call

| | | |
|---|---|---|
| average capacity cost ratio | = £34M/£60M | = 57% |
| average gross profit ratio | = £6M/£60M | = 10% |
| contribution ratio | = 57% + 10% | = 67% |

The contribution ratio is the average contribution (i.e. sales minus variable unit costs) needed from sales in order adequately to cover the fixed capacity costs throughout the year and the 10% profitability requirement.

Now, suppose he wants to do some business with the Zeta system. This system is old and no longer actively marketed, but a customer has decided to install a system at one of their subsidiaries. The customer is willing to pay £9000 and Leonard has a full-cost calculation from the accounts department of £11,000.

As Leonard knows that the Zeta system needs very little customisation, no marketing or selling, very little administration and is very easy to put together, he is certain that the capacity ratio will be only around half its normal value, i.e. at most 30%. This value is reasonable as when he adds the 10% gross profit ratio he will arrive at a contribution ratio of 40%, which generally gives fair profitability.

Leonard also knows that the variable cost for Zeta is £3500. As we have found that the average contribution ratio for Zeta should be around 40%, the variable unit cost will be about 60%. He concludes that £3500/0.6 = £5833 is the price that would give a normal profit level. So £9000 would therefore be excellent business.

This proved to be true. After checking up on the calculation methods used, it turned out that some old figures had been used to distribute full administration, sales and development costs to system Zeta.

## *Calculate costs, not prices*

The risk that Leonard was running was that this calculation method can be a backdoor way into cost-based pricing.

However, when pressed for time, it is easy and useful to multiply unit costs by an 'approved' factor to arrive at a safe price.

The fact that the customer could have been willing to pay twice that price is quite another factor, one which would not be realised unless you know the market well and spend sufficient time and effort exploring and developing all the possibilities.

# Summary

A post-industrial company cannot use full-cost calculations to prioritise business opportunities. A well thought-through opportunity-cost calculation should be used.

To start with, it is important to determine the situation in which we are calculating. There are six different, alternative scenarios and we must take all six into consideration; whether we have over- or undercapacity in the short- or in the long term and, for long-term decisions, do we have a red or a green light?

1. In the short term, it is important to use any overcapacity we may have, but not at any price. Before taking business at a lower margin, every other alternative should have been tried. An increase in sales will require extra marketing activity, which contradicts the old wisdom that 'if you do not reach your income budget, you must not exceed your cost budget'. Another problem is that sound financial strength is required to use overcapacity for activities other than low-margin business,  such as development and training.

2. In times of undercapacity it is tempting to increase prices in the short-term as we have more orders than we can handle anyway. A pricing policy that is too opportunistic, however, could make it difficult to attain our long-term objectives — namely taking and defending a dominant share of our chosen niche market.

3. Undercapacity normally stems from a constriction somewhere in our capacity. Of course we want to invest in order to remove that constriction but if, as a result of this bottleneck, we already have a red light for profitability and financial strength, the board may not agree to our investment. In that case, we have to try other ways such as buying in capacity etc. If this is not successful our only alternative is to shrink the organisation to achieve balance in our capacity.

4. If we have a green light, we can go ahead and invest to remove the constriction in our capacity.

5. In a situation with a weak balance sheet and overcapacity, it is tempting to try to expand out of profitability problems. Unfortunately this can worsen the situation further as expansion requires additional investments in market and system development, which will have to be written off in one year. It also means increased capital tied up in marketing assets. A better way might be to increase prices and reduce the size of the organisation.

6. A strong balance sheet and overcapacity is the ideal situation to attack our competitors' market shares. We must, though, resist the temptation of running in all directions at once. **We must stick to the straight and narrow.**

# Part II

# DEVELOPMENT

# Chapter 8

# System development

*In the industrial society, we developed relatively simple products. In the post-industrial or information society, we develop systems characterised by a functional combination of hardware and software together with what can be called documentware.*

*We have always had to create a perceived value, for which the customer is prepared to pay. As the full value of post-industrial systems is seldom visible to the naked eye, communication of a system's potential value to a customer has to be developed in parallel with the technology itself.*

Many books have been written on the technological aspects of product development, so I shall not try to compete with them. What I hope to achieve in this chapter is to widen the concept of product development and to develop the commercial aspects.

I want, for example, to stress the importance of looking at the pricing of a system as soon as the first idea of a new concept is formulated.

I was once part of a management team discussing a new system under development. When I asked, for the fifth time, how this system was going to be priced, the MD lost his composure and shouted, 'For heaven's sake, we can't start discussing pricing now! We still don't know if we have anything to sell!'

By the time a system is developed and ready to sell, it can be too late to think of how to price it; the system may already have been constructed in such a way that it can only be altered with difficulty.

# System = product + service

*The word 'product' makes you think of hardware*

The first major industry that had reason to stress the value of intangible goods was the computer industry. Terms like hardware, for machines, and software, for computer programs, were coined.

## Hardware

Hardware in its simplest form has the advantage of speaking for itself. It can be beautiful or ugly, light or heavy, rigid or flexible.

Much of what we say concerning economic and business development assumes we are dealing with hardware which has easily measurable attributes. Weighing up price against quality is much easier for standardised hardware than for a service.

## Software

Because the computer industry was the first to use the term software, to many people this word has become synonymous with computer programs. The concept can, however, usefully be widened to describe other intangible aspects of a system, such as user training, system installation or maintenance.

## Documentware

As computer software is itself rather abstract, it has to have a visible element or interface in order to be useful. A function which is unknown or unintelligible to a user is of little or no value to the user.

One computer manufacturer highlighted a function in its personal computer which it called the 'secrecy button'. By using this button, it would be possible for a user to prevent unauthorised users accessing the hard disk. In reality, anyone who understood the operating system — which was a well-known industry standard — would know that this function was available as standard on all personal computers. However, my friend, who knew little about computers, chose and bought this particular brand because he wanted this function, and no other manufacturers had informed him that their machines also featured it.

Nowadays the importance of the documentation accompanying a system is recognised and one often talks of three parts of a system: hardware, software and documentware.

This is not to say that we need to supply users with tons of paper every time we ship out a system. If a computer is part of the delivery, it is often advantageous to incorporate the documentware within the software in the form of a 'help', tutorial or information program. Perhaps the term 'user help' is more appropriate than 'documentware'.

## Component or system sales?

As the proportion of hardware is drastically reduced in our modern systems, we may have problems in that customers will not always be able to 'see' what they are paying for.

In the first chapter of this book I related a story about a telecommunications authority receiving tons of hardware when they ordered an expensive new telephone exchange several decades ago.

If you bought an add-on function to your digital exchange for an equivalent sum of money today, you might just receive a visit from an engineer to install some new software. There are still some old-timers in the authority who would consider this to be 'after-sales service' and that it should be supplied free-of-charge.

As the marketing of products and services may have fundamental differences, companies sometimes decide to market the two components separately. The system side is sometimes considered as a separate consultancy, which charges separately for its services.

One advantage of such a split in business is that no-one can now demand to receive the consultancy work for free. A disadvantage is having to sell unique knowledge on a 'per hour' basis.

You may remember from Chapter 4 that OPERATOR Ltd. had good, as well as bad, experiences from both ways of working.

## Small systems or large?

Having overall responsibility for a system can be both positive and negative. One advantage is that you can offer a competitive and complete system (which some might call a 'one-stop' solution) to a customer keen to avoid the trouble of having to deal with multiple suppliers. For larger systems, there will also be more possibilities for flexibility in pricing.

However, if the prices of the individual components and sub-systems are well known, such possibilities for creative pricing are greatly reduced and you may be left with only pricing your own knowledge in integrating the components. The more sophisticated the customer, and the better known the prices of the components and sub-systems, the less the customer is willing to pay for your knowledge.

The most difficult situation is when our overall system responsibility includes responsibility for other suppliers' sub-systems. We normally do have ways to obtain redress in case of problems with their components, but proving this can become costly. To get really effective redress, our subcontractor must have the financial ability to pay guarantees.

Total system responsibility is most successful when all components and sub-systems are produced in-house.

Even if you decide to minimise system responsibility by only selling components, it is often impossible to limit oneself to supplying only hardware. The demands of customer-specific products are increasing and a certain amount of documentation and training will often be required along with our hardware.

The question of whether to supply components or systems may have been posed incorrectly. Perhaps we should rather ask whether to deliver small or large systems. The answer to that question depends on what we consider to be strategic added value.

## *Strategic added value*

In Chapter 4 I encouraged you to 'stick to your knitting'. International competition compels us to try to become world leaders in our own speciality; even if we do not plan to compete overseas, we have to be resigned to the fact that we will meet international competition in our home market.

The more narrowly we define our speciality, the easier it is to become the best in the world.

In the first chapter, we discussed the pros and cons of a high added value. We found that the drawbacks can outweigh the advantages. Every opportunity should be taken to 'farm out' whatever we do not necessarily need to do ourselves. This makes sense for two reasons; it increases our cost flexibility and gives us the opportunity to concentrate all our resources on what we consider to be our strategic competence.

I think development today is moving towards making system responsibility and product management a business idea on its own.

The ingredients for success will be a profound knowledge of customer problems, combined with total freedom to exploit all the various components and sub-systems available in the market.

For those manufacturers who do not have system responsibility, the challenge is to develop competitive components and sub-systems with clearly defined interfaces that easily fit the system-builder's requirements.

The choice between the role of system integrator and subcontractor is a major strategic decision.

## Strategic alliances

The customer's requirement for a total system that functions well may not accord with our own wish to base our competitiveness on well-specified niche products.

One way to satisfy the customer will be to form strategic alliances with other companies marketing complementary components and systems.

Such strategic alliances can be anything from a subcontractor agreement, with a minority shareholding and exclusive rights, to a verbal agreement simply to recommend each other's products or services whenever meeting customers with requirements we cannot fully satisfy ourselves.

## Your comments

*Does your company supply large or small systems, or has it managed to find a niche where it only supplies components? Do you have a well thought-through strategy on what you consider to be your strategic added value or do you try to bite off more than you can chew?*

# How to create customer value

## *Market- or technology-driven system development?*

In post-industrial business development, we need both technology- and market-driven system development. The two are related rather like a see-saw, where movement one way leads to the other. Technical break-throughs sometimes make it possible to solve problems hitherto considered insoluble.

The trend, however, is towards taking the market as a starting point for system development, as we today have so many technologies and technological resources that we can solve most problems by simply applying what is already known.

This should not stop us from investing part of our development budget in basic research. It is, after all, the technological break-throughs that give the real advantages in competitiveness. However, it goes without saying that all commercial activity demands that there be a customer problem at the end, solution of which will make it worth our while.

## *The customer's problem — is it worth our while to solve it?*

All applied system developments should start by defining the customers' problems and assessing what the solutions could be worth to them.

In principle, you can create customer value in three ways:

1.  Increasing the customer's income;
2.  Reducing the customer's costs; or
3.  Increasing the customer's security.

In order to illustrate this, I shall refer again to our example company OPERATOR Ltd. As before, understanding the detail of the technology is irrelevant; it is the overall business picture and methodology that are important.

When OPERATOR started developing a new semi-linear system, development was based both on a known demand in the market and a new break-through in technology.

The technical break-through was not made by OPERATOR but by its suppliers of sensors and processors. With new hyper-sensitive sensors and high-speed processors, OPERATOR could tackle the real-time control of combustion processes in foundries, which had not been possible before.

Energy and wastage represent large costs to foundries. If the new semi-linear system could be given the right properties, it should be possible to reduce both wastage and energy consumption drastically. In total, there could be annual savings of half to one million pounds for each customer.

In general though, the largest profits tend to come by increasing a customer's income. By improving flame control, OPERATOR claimed that it was possible to see an improvement in quality which, if proven, could lead to the foundry being able to increase its prices substantially, at least in the more quality-conscious sectors of the market. This could, however, be a rather difficult argument to carry through and quantify precisely.

The development engineers in OPERATOR were convinced that safety in foundries could also be improved by better flame control, as well as the reduced costs of wastage that had already been proved. Wastage also tends to mean delivery delays which can reduce a foundry's reputation as a reliable supplier of 'just-in-time' products.

However, as it was only the cost savings that could easily be demonstrated and proven, OPERATOR decided to base its pricing on a customer value of between £0.5 and £1.0 million per year.

With a little more patience and careful development of the arguments, this figure could possibly have been doubled or trebled, but it was considered to be wasted effort. The attitude was one of 'we can't charge for it because we can't explain exactly what the customer is paying for'.

## *Our solution — what is it worth to us?*

We now know more or less what a successful solution could be worth to the customer, but what is it worth to us? How much of the customer value should we get?

The answer depends on our negotiating strength and our ability to communicate. One way to set a price would be to share the value of the solution with the customer. In this example at OPERATOR we could probably get about £1.0 million per customer, provided we calculated on the basis of five years of income surpluses in today's money.

But a million pounds for what exactly?

## *How do we make it tangible?*

A very common problem in post-industrial business development is that we sell invisible values. How can we expect the customers to understand what they are getting for their £1 million?

> OPERATOR was prepared to reduce the price of the semi-linear system compared with the old linear system. The old system required a minicomputer whereas it was quite possible to run the new system in a PC environment. There were long discussions in the development department on whether to develop such a slimmed-down version or whether to base a new system on a larger computer.
>
> Once the problems with pricing had been recognised the decision was to develop the system around a larger computer.

It was a sensible decision; had the development been made on a PC, it would have been very difficult to break through the psychological price barriers perceived by customers. Nobody feels they should pay a million pounds for a (standalone) PC-based system!

## *What does the system development cost?*

Having arrived at a pricing potential of well over a million pounds per customer, all OPERATOR's doubts about the eventual profitability of the project were gone. What remained was to calculate how much its financial strength could stand in the way of development costs.

Charles Armstrong and his Finance Director found that it would be possible to invest at least half a million in the current year — a figure well in excess of the amount required by the development department — so Charles gave the project a green light.

As he is basically a 'belt and braces' man, Charles initially only allocated money for a market study and a technical pre-study.

### *Stop/go test no. 1*

Charles Armstrong had been in the business long enough to realise that the last thing he wanted was his own people to carry out the marketing study. Enthusiastic people only see what they want to see.

Once the independent market study had found a high customer acceptance for the new solution, and the technical pre-study had confirmed that no insurmountable technical obstacles existed, Charles gave the go-ahead to develop a prototype.

# How to make the customer perceive the value

## *Communication of what may not be obvious*

As we have noted previously, much of the value of our business offerings is no longer in the material items or hardware but in less tangible properties. If much of the value of all our development effort is hidden, we must spend extra time and effort on communication and information. What could be more natural in the information society?

I remember being part of a ten-strong development group discussing a new system the company wanted to develop. Half were European and the remainder North American. In retrospect I think there were as many different views of our development as there were participants.

Communication of abstract values is always difficult — none of the group was a communication professional but two of the Americans were more skilled than the rest in this area.

The main task of the system under discussion was to measure and describe small particles, but the group was split over the question of the capacity needed for the system. The British thought the Americans had taken leave of their senses when they mentioned how large their requirements were.

To the surprise of the other participants, one of the Americans brought a half-empty salad bowl into the conference room after lunch. He then asked to have the sugar bowl, from which he took some lumps of sugar, placing them in the salad bowl and giving it a good shake. He asked the Englishman next to him to close his eyes and take a sample, first from the sugar bowl and then from the salad bowl.

Not surprisingly, the Englishman picked up a lump of sugar from the sugar bowl and a piece of lettuce from the salad bowl. The salad bowl represented the American market, the American explained, and the sugar bowl the British. To arrive at the right results in the USA, samples have to be that much larger than in Britain.

Instead of putting on this show, he could of course simply have said that the particle size standard deviations in the American market are bigger than those in the British, but somehow I don't think this would have made the same impact.

What I want to get across is that a communication professional should be present right from the first development meeting, whether from an advertising agency, your own commercial department or — at the very least — simply somebody with a flair for stories and drawings rather than dry minutes and long-winded reports.

I would even go so far as to question the sense in investing money in development at all if you do not have the resources to communicate it. In the long run, professionals are often less expensive than amateurs.

## Stop/go test no. 2: sales training after prototyping

It is of course vital that we can explain to ourselves what we want to achieve, but that is not enough. As we are involved in commercial activities, we should involve the buyer as soon as possible in this process.

In consultancy, customer value is completely abstract and it is generally true that development in this industry is best when it's being paid for.

This is not being cynical but simply an expression of the belief that the presence of a customer gives stability to the development process and focuses it in a way not possible with entirely in-house development. There is a tendency in some small, high-technology companies, when they obtain government money to start independent development, to squander resources on  technology that interests them but that no customer is ever going to want or pay for.

Involvement of a customer in development from the outset may not always be realistic, but perhaps we can learn something in this respect from the consultancy industry. In applied system development, the customer should be involved in the complete development process from the first marketing survey to the last sales training session prior to launch.

If, for practical reasons, the customer cannot be present, he or she should be represented by someone close to them; for example, our sales force. Ideally, though, customer interest should be represented by someone external to the company who is not afraid of expressing uncomfortable views.

As an example, there may be people available for paid consultation who have been in line-management functions with our customers and who often remain in regular contact with them.

I am aware of the problems we face with confidentiality when we use external resources but that has to be resolved somehow. I am sure you have heard of the development engineer with the sign on the wall behind him saying 'my job is so secret that I don't know what I'm doing'.

It might be appropriate to mention that it is easier to take legal action against a company breaking a confidentiality agreement than against an employee doing the same thing.

We should arrange to have sales training as early as possible, where our offer is presented to a real or imagined customer at the highest price level we consider defensible.

But what do we mean by 'as early as possible'? I really mean the earlier the better — at the very latest when we have a prototype.

At such a mock customer presentation, not only can we confirm the customer value but, in the ideal case, we will also gain more ideas for improvements — making the system even more valuable than we had perhaps imagined at the outset.

In the worst case we may realise that the customer value is so low that we must stop our development, in which case the earlier this happens the more resources we will have saved.

## Sales training before pricing

When the system is complete, it is time to check the 'communication package'.

We have taken the important strategic decision to price according to customer-perceived value (CPV). If we only communicate 30% of the customer value, we are only likely to be paid half of what we could get had we communicated 60%. Achieving 100% is probably unrealistic.

Considering the importance of communication in getting paid for abstract advantages, our communications should be tested as thoroughly as the technical system. Once we have the first draft of all the planned communication material, it is time to test the communication of advantages and profits in as realistic a sales environment as possible. Not until then can we finalise our communication package and thereby our pricing.

What about calculations?

Naturally we do have to check the profitability of the project; we can do that continuously using the principles we learnt in Chapter 7.

# Your comments

*How does development of communication of customer value in new systems take place in your company?*

# How to retain the value

The post-industrial or information society is characterised by rapid communication. Where we could previously expect to develop our business in the home market at our leisure before overseas competitors became a threat, we now have to accept that news of our systems will travel fast and that our competitors will do their best to copy our success.

Once there is an alternative to our system, the base for our pricing will shrink dramatically. We can then only ask to be paid for the additional value we create compared with alternative solutions to the same problem.

Naturally we want to make it difficult for others to profit from our innovation.

## *The first line of defence is a patent*

A patent is the traditional way to protect an inventor's right (of time) to exploit his or her invention without competition. Patent protection is what we desire most of all; every possibility should be tried to acquire a patent.

There are sceptics who say that there is no point in applying for a patent because competitors will find a way round it. It is a fact that patent registration is an efficient way of telling our competitors that we have found something useful.

It has been claimed that one Japanese patent strategy is to create a ring of application patents around a foreign basic patent. By doing so, opportunities for the original inventor to use the patent will be limited.

This view should not be used to prove the futility of applying for a patent. On the contrary, the conclusion would be to stress the importance of applying for more than one patent, in order to block competitors trying to circumvent an individual patent.

The trend towards more applied development does mean increased patent costs; it is costly to apply for a patent and defending one is likely to cost even more.

Pessimists may draw attention to the fact that some countries do not prohibit the party accused of patent violation from continuing to use it during the trial process. This means that the accused party may well file a countersuit in order to try the validity of the patent, which can take quite some time. From the start of the original trial, several years can elapse before any real conclusion is reached, thus giving the guilty party plenty of time to compete.

But, as I am an optimist, I still think that patent registration is a good investment, although it is very important to have sufficient financial strength to be able to carry it through all the way. As I have mentioned before, a good summary of patent and other intellectual property rights can be found in *'Intellectual property rights for engineers'* by Vivien Irish[1].

## *Our second defence is protection against theft*

There are other possibilities for improving protection. Safeguards can be built into systems and programs and legislation other than patent registration can also be used, such as copyright protection, registration of a trade-mark or brand name and so on.

## *The third measure is 'positioning'*

Another way to achieve an advantage over future competitors is to acquire the 'No. 1 Position'. This can normally be achieved by quickly taking a large share of the market. The first company to launch a product in the market and acquire significant sales inspires a natural confidence compared with latecomers selling copies of the original.

It is important to have sufficient strength to capture a large market share in the initial phase of business. If someone else can overtake us, the market will soon forget who was first.

---

[1]Irish, V., *'Intellectual property rights for engineers'*, Institution of Electrical Engineers, 1994.

To be overtaken is exactly the risk Leonard Hewson now runs in OPERATOR.

It is too much to ask that you remember the first page of this book, now that we have come this far, but there we found Leonard Hewson bitterly complaining that OPERATOR, due to the recession, did not have enough strength in its balance sheet to fulfil the launch plans for the semi-linear flame control system. Leonard claims that spending money on system development without having the funds to launch it into the market properly is like building a house without being able to afford the roof.

Just as important as selling in quantity, and doing it quickly, is to supply good quality. To be known as a low-quality supplier will not help us at all. It is important to be able to guarantee adequate quality. This can seem obvious, but I have seen many examples of companies being so caught up in the rush to be first in the marketplace that quality has been sacrificed.

## *Guarantee costs — a guarantee for quality*

The more customer-specific and the larger our deliveries, the greater is the risk that we will have to make adjustments and give additional service.

Legislation in most countries compels us to assume a guarantee responsibility, so why not take the bull by the horns and have a proactive guarantee? By a proactive guarantee I mean a guarantee which clearly communicates the customer's rights according to the legislation of the country in question and, additionally, includes any necessary services and post-adjustments to ensure satisfactory functionality.

Theft and piracy problems can also be reduced via the guarantee. Any unauthorised copies, adjustments, components etc. will make the guarantee null and void.

The duration of the guarantee has to be limited, if for no other reason than to enable us to reserve the necessary funds. Most tax authorities allow us to set aside funds in order to allow us to meet current guarantee obligations.

It goes without saying that, in our economic planning, we have to set aside resources to be able to fulfil the obligations in our guarantees. This can also be considered to be an investment in image.

Again I would like to stress the importance of a well thought-through geographical strategy. With extensive guarantee commitments, it may not be wise to have them spread right across the world.

# Summary

In the post-industrial or information society, product demand becomes more abstract. Less and less hardware in isolation, and more and more combinations of hardware, software and documentware are required. Let's call that combination a 'system'.

A question of great strategic importance is whether to supply large or small systems. If the systems are limited when it comes to software and service, we could call them 'components'.

When **development** becomes more important than low prices, we have to set our price according to the customer's perceived value (CPV) to ensure maximum resources for continued development.

Development work has to be concentrated in two areas; first, creating customer value through technical development and, second, efficient communication of this value. Both must be developed in parallel as communication is an integral part of customer value and crucial to our ability to be paid for the solution.

Value for customers can be achieved in three ways. You can increase their income, you can decrease their costs or you can increase their security, which must then be converted back into increased income or reduced costs.

Once it is clear that we can achieve a technical solution to generate customer value, it is important to assess this value in financial terms. How we and the customer share this value is a matter of negotiation and communication. The result of this process will decide the future resources available for future development.

A successful solution will quickly be copied by competitors if nothing stops them. Primarily, we should try to acquire a patent but we also try to minimise our risks by building copy protection and other features into our systems. By quickly establishing ourselves as number one in the marketplace, we will achieve a strong position in the long term.

Just as it is important to test the technology in a new system, it is also vital to test our skill in communicating the advantages and profits available from the system. This is carried out through repeated sales training sessions.

Chapter 9

# Price development

*In the industrial society we were ruled by external price development.*
*In contrast, in the post-industrial society* **we** *develop the prices.*

Most people consider 'price development' to be something harmful to
the company:

'With the price development we've seen in the market over the last
year, we will hardly be able to improve our result at all.'

This view was natural, and perhaps even realistic, in the industrial
economy. When you sell standard products, with sufficiently long life-
cycles, competition will drive prices down to the level where only the
most cost-effective companies survive.

When you sell sophisticated systems in the post-industrial economy,
development costs will drive down profitability to the level where only
the most revenue- and cost-effective companies survive.

**Now we have to develop our pricing in order to maximise our**
**revenue** in order to be able to invest in further development.

## How to develop our pricing

*We get the price we can justify*

When selling systems with invisible values, we can only get the price
we can justify. At the risk of sounding like a broken record, I repeat that

if we can only communicate 30% of the advantages and profits that our system can give the customer, we can only charge half as much as if we were communicating 60%.

It is not justifiable to charge high prices simply because we have high costs. We only justify charging higher prices if we are giving the customer an even higher value or benefit.

## *How much should we charge?*

How much we should charge depends on three things:

1. The value of our offering, relative to the customer's other alternatives;
2. How well we are able to communicate our advantages;
3. Our creativity and negotiating skills.

The first two points were covered in the last chapter; now we will concentrate on the third.

Take OPERATOR's new semi-linear flame control systems, mentioned in the last chapter. Let's assume we succeed in communicating that the value to the customer is a saving of £1.0 million per year over his or her alternative if it is a manual control system. If they already have our linear flame control system, the value of the saving will be £0.5 million, or £0.6 million if they have our competitor's linear system.

How much can we now charge for OPERATOR's new semi-linear flame control system 'Newfound II'?

All the data are listed in the following table (but, again, do not worry about the specific numbers or the technology):

| COSTS per system | 000s Variable | Full cost |
|---|---|---|
| Hardware (computers & sensors) | 40 | 70 |
| Software and documentation | 0.3 | 30 |
| Solvent / year | 80 | 88 |
| Consumables / year | 1.0 | 1.1 |
| Service / year | 0.8 | 1.2 |
| MARKET POTENTIAL (no.of systems) | Europe | N.America |
| Our old system | 100 | 50 |
| Competing systems | 400 | 500 |
| Manual systems | 900 | 800 |

The full-cost calculations are of course very approximate. They are based on sales forecasts squeezed out of a reluctant Leonard Hewson by Malcolm, combined with historical cost distribution keys.

The market potential is just as approximate, but you can assume that the numbers are on the low side. Leonard is always careful in his judgement as he fears that one day 'the numbers will be used against him'. It is an interesting fact that the larger foundries will need more than one system.

Further to the above, I should mention that our patent registration has just been approved.

# Room for thought

*How would you draft a pricing method for OPERATOR's new flame control systems? If you are not sure, read on.*

The answer to how OPERATOR should price its new semi-linear flame control systems depends, among other things, on how it will charge for them.

## How should we charge?

The way in which we charge has a clear impact on how much of the customer value we can persuade the customer to 're-invest in continued development'. That is to say — how much we can persuade them to pay us.

I can list five ways of charging :

1. **All the money up-front;**
2. **Monthly payment based on usage;**
3. **Payment for usage per unit;**
4. **Payment for consumable items;**
5. **Profit- or risk-sharing.**

I am sure that there are more besides those listed and only your own creativity will limit you in deriving more. Let me now comment on my five ways, one at a time.

### 1. All the money up-front

This is liable to make life difficult when selling invisible values, but the effect on our cash flow will be tremendous if we succeed.

One problem in asking for payment up-front is that the buyer's decision-making process may be much longer. The purchase will be considered as an investment and, if it is of a reasonable size, it may have to be brought up at a board meeting — and a board will not approve an investment without an investment calculation.

Investment calculations are often overlooked as a sales tool.

Too often a supplier leaves the task of constructing investment calculations to the buyer, which is a lost opportunity. It is, after all, through such investment calculations, among other things, that we justify our price.

The question now is what type of calculation to use.

As we saw in Chapter 7, most investment decisions are made based only on a simple pay-back calculation:

$$\text{PAY-BACK TIME} = \frac{\text{INVESTMENT}}{\text{INCOME SURPLUS}}$$

This is an investment calculation you can be sure that the whole board will understand, so let's start with this one. You do not have to be a mathematical genius to rearrange the equation above to read:

INCOME SURPLUS  x  PAY-BACK TIME  =  max. INVESTMENT

Returning to our example, if we use the least cost-saving alternative — which we know to be £0.5 million per year — and deduct the cost of solvent and other consumables, we will have an income surplus of around £0.4 million per year, so we can fill in the following table:

| INCOME SURPLUS | x  PAY-BACK TIME | =  max. INVESTMENT |
|---|---|---|
| £0.4 M | 1 year | £0.4 M |
| £0.4 M | 2 years | £0.8 M |
| £0.4 M | 3 years | £1.2 M |
| £0.4 M | 4 years | £1.6 M |
| £0.4 M | 5 years | £2.0 M |
| £0.4 M | 6 years | £2.4 M |

The price will then be the maximum investment sum we can defend in terms of pay-back time.

# Room for thought

*What would your price be?*

I think pricing OPERATOR's system at £0.4 million — that is to say the investment would be paid back in a year — would be too easy.

A six-year pay-back time is the longest I have ever been able to defend; that took several weeks of hard work together with the best marketeers in the company and their advertising agency all struggling to find sufficiently strong arguments.

The risk, of course, is that someone with a financial background will insist upon a more complex calculation, taking interest into account. This is not difficult to achieve provided you make sure you compensate the income surpluses for inflation.

One question you must be prepared to answer is how the pay-back time relates to the lifetime of the system.

In the case of OPERATOR, we can guarantee a lifetime of fifteen years. We can therefore argue, based on annual savings of £0.4 million:

'If you buy our new system, Newfound II, for £2.0 million, it will be paid back in five years, after which it will earn you £0.4 million per year for an additional ten years.'

'But that's nonsense,' the customer's Finance Director might argue. 'You cannot calculate using a technical lifespan. Anybody knows that you should use the economic lifespan. This system is mostly electronics; an investment in computers or electronics where we can't have pay-back in under a year is of no interest to us.'

This suggests the most that customers would readily pay is £390,000.

# Room for thought

*How would you answer that Finance Director?*

If you cannot find a good answer, the price per system would come down to this figure of £0.39 million rather than the £2.0 million OPERATOR was initially seeking.

This is what I mean by 'getting the price we can justify'. When the arguments run out the price goes down. Instead of just looking at our cost analysis we need to spend time and money developing our arguments and justifications.

---

## Rooom for thought

*Do you have any more ideas?*

It isn't easy to counter that Finance Director's argument, but you could always try something along the lines of:

'You are correct in saying that the pay-back time is most often compared with the economic lifespan rather than the technical, but that is when we are talking about uncertain income surpluses. In your case it's a question of saving energy as well as reducing wastage — that will always be important, for as long as you have a foundry business, won't it?'

'That may be the case,' the Finance Director might answer. 'But that's not what I'm talking about. I believe that in a year from now someone else will offer us a system that will save us £1 million a year and then your system will be obsolete.'

## Room for thought

*Now what do you say?*

I would counter like this:

> 'You're right — the industry is constantly developing. But our company is leading it and since we have patented this system, we will be the ones who will approach you with an attractive proposal for an update. However, irrespective of who it is, you'll demand sufficient increased savings from a new system that will make it pay to scrap this one.'

You could then back it up with:

> 'And I'm sure you'll agree that it isn't sensible to wait until development has stopped before you buy a system.'

That is one way you could develop the sales argument, but it is important to have thought it out and have it entirely clear before meeting the first customer.

In general we spend far too little time on developing these pricing and value arguments before launching new systems.

As I have said before: 'if you can only communicate 30% of the advantages and profits your offering can give the customer, you will only be paid half the amount that you could get if you communicated 60%'.

> Suppose we decide that it is possible to defend a three year pay-back time for that system, and that we can therefore fix the price at £1.2 million per system. When the customers have the system installed they'll probably threaten to call the police (or the Office of Fair Trading):
>
> '£1.2 million for a little box and a program!'

You do not make life easy for your sales people if you ask for all the money up-front when selling invisible values. It could be much easier to ask for payments spread over time.

## 2. A monthly fee

When presenting this alternative, it is important <u>not</u> to make it sound like a leasing deal as you would then run the risk of having it treated as capital investment anyway.

In the case of OPERATOR, I would recommend an agreement including as much service as possible.

> 'We'd like to offer a full service agreement to reduce your energy and wastage costs by about £0.5 million per year. We will construct, install and manage the flame control system and all necessary equipment for £24,000 per month. Training, documentation, performance control, monthly service and everything needed to run the system, such as consumables and solvents, are all included. And we will guarantee the result!'

The drawback of this method is not having control over the use of consumables. It may not matter that much in this case, but in other situations it could be disastrous to give responsibility for using the consumables to someone who is not responsible for paying for them. A step in the right direction could be to ask for a results-oriented fee.

## 3. Payment per unit produced

> For OPERATOR, this would probably mean having to tie the price to the amount of metal passing through the foundry. This would take into consideration variations in utilisation. If the customer increases the production by one shift, the usage of solvent and other consumables will increase but so will our income.

Unfortunately we have still not eliminated the risk of wasteful use of consumables. We can reduce the risk to some extent by supervision and control; we could also build in a discount system to reward low usage.

Alternatively, why not eliminate the problem completely by asking for payment based on consumption?

## 4. Payment for consumables

Based on the assumption of annual savings of £400,000, we could price consumables and service in such a way that the price per year would be £380,000. With a price for hardware and software of £80,000 we could present the proposition in the following way.

'We offer you a very powerful minicomputer system with special programs for £80,000. Using this system, we guarantee you a reduction in energy and wastage costs of £500,000. To run the system, you need a special solvent, certain other consumables and on-going service. The annual cost for this package will be a maximum of £380,000. Your profit on the system will therefore be £120,000 per year which means you will recoup your investment in just eight months.'

There are drawbacks with this pricing method. Immediately the customer receives the first shipment, he or she will instruct the chemists to analyse it.

'This is common kerosene at triple the price!' is what they would have said, feeling cheated, if Brains had had his way.

When the solvent was developed, a special type of kerosene had been found, but it turned out that the results were only marginally better than those with normal kerosene. The development manager (Brains) therefore decided to use standard kerosene, allowing the customer to buy it him- or herself and saving us the trouble of handling the product.

Fortunately, however, the marketing department had become aware of the pricing problems and had managed to convince Brains it was worth the time and effort to use the special version. They also persuaded him to add certain tracer elements to help reveal attempts at copying it because, naturally, all our guarantees would become null and void if the original fluid is not used.

The second weakness of letting the consumables carry such a large proportion of the price is that we do not always know what the cost savings will be in reality. Energy prices do vary.

Finally, we come to my last alternative.

## 5. Profit/risk sharing

We could imagine proposing the following offer to OPERATOR's customer:

'We will install our new flame control system, NewFound II, free-of-charge. When the savings become evident and quantifiable, we will split them with you fifty-fifty.'

I would not recommend OPERATOR to use this charging method but it has been usefully implemented in many industries. In the early days of telecommunications, hardware suppliers sometimes operated the entire telephone system in a country and then shared the profits with the government of the host nation.

## Use a combination of methods

My recommendation would be to use a combination of payment methods.

Some guidance can come from the computer industry, where you often see many skilful combinations — say a cash payment for hardware, a monthly fee for software and a service agreement to have the functionality guaranteed.

## How to explain why we charge so much

However advantageous our offer may be, and however we present our price, we will be questioned.

We just have to accept the fact that our price will be questioned, irrespective of whether the calculations have been based on costs or fixed by skilfully finding the customer-perceived value. One of the professional buyers' most important tasks is to question what they are getting for their money.

What we cannot accept is our own doubt about our prices. As we have said previously, if we are not entirely comfortable with the price ourselves, our body language is likely to reveal our doubts. If you do not recollect our discussion of morality in pricing and the importance of countering one's own objections, you should re-read the section towards the end of Chapter 5.

Once you realise that the customer often claims more than half of the customer value, without making any contribution other than 'owning the problem', there is a risk of becoming arrogant.

'It's none of your business what it costs me to produce this system — you'll make £120,000 per year without doing anything but housing it!'

How much you will have to explain is entirely a question of your negotiating position and your ability in communication. I will give you two examples; one where the customer value was proven beyond doubt and the seller had a very strong position, the other where it was much less certain and the seller was completely dominated by a strong buyer.

My first example concerns a tax consultant who rang up a local company and said:

'I read in the paper that you have sold your industrial property in Warrington. I have calculated that it will cost you some five hundred thousand pounds in tax — I can reduce that to one hundred thousand.'

'That sounds interesting — what would you charge?'

'Nothing if I'm unsuccessful, but I want ten thousand if I succeed.'

The tax consultant was successful and his client received an invoice reading simply: 'FEE — £10,000'. A note was attached to the invoice saying: 'If you ask for a breakdown of my costs, I won't help you again'.

My second example is about an engineering consultant working for a government department. The buyer was very knowledgeable and ran a development project where some of the capacity was bought in from independent consultancies. The negotiations were carried out according to a set pattern and the consultant was so tied up with clauses guaranteeing his customer access to internal information, that the buyer could call up an engineer and ask:

'What's your salary?'

'Why do you want to know?'

'I want to check the mark-up on my latest invoice.'

It has even been known for a consultancy to be forced to issue a credit note to a buyer because a younger engineer with a lower salary has been used for a task after receiving instructions from an older colleague.

These are two extreme examples showing that we live in a market economy where there is competition and free negotiating rights. However, most of us are happy to live like this, knowing that bureaucratic control is usually even less effective.

## Your comments

*Where on the scale between the independent tax consultant and the dependent engineering consultant does your company fall? Can you do anything to increase your independence?*

Most of us fall somewhere between the tax consultant and the engineering consultant in my examples and, irrespective of how favourable our offer might be to the customer, we will still have to explain why we cannot make it even more so by reducing our price.

'It's for your own good — if you force me to reduce the price further, I will not be able to come back in the future with even more attractive offers.'

Sadly, an argument like that will not convince the customer. Customers cannot understand their own good; a pound today is often worth more than ten tomorrow — and, anyway, why not let someone else pay for the development costs? It won't be more expensive just because my company can enjoy the results.

Therefore we need to keep a straight face and patiently explain our prices, even if we may feel we shouldn't have to.

The paradox we face is that we often have to make it sound as though we have used cost-based pricing, since the customers will not rest until they think they have pushed us down to a price that just about covers our running costs and leaves no margin for bold development projects.

## The problem of excessive profits

One of the biggest challenges in post-industrial business development is to retain the resources necessary for long-term development.

If, in the interest of our customers, we can retain a substantial proportion of the system value, with the present system of accounting this will show up as 'excessive profit', something everyone wants to get their hands on — society, employees, customers, suppliers and shareholders alike.

## Profit fluctuations will increase

With a higher proportion of fixed capacity costs, fluctuations in the economic cycle will have a greater impact on our profitability.

The increasing development investments, being written off in one year, will also contribute to the variations. If, in addition, you consider that only one in three development projects is commercially successful, you realise the impossibility of judging a development-intensive company's real profitability over a single year.

We need many profitable years to compensate for lean years when we are developing. These companies must be allowed to build up a strong balance sheet with reserves to be used during the leaner years.

## *There are no excessive profits*

As the strengthening of the balance sheet results in an increase in the shareholders' capital, naturally they will consider withdrawing some or much of it to increase their own wealth.

But what seems to interest the shareholder most is business, so the surplus will often be re-invested in business development. In the long term, profitability will tend to stabilise around the base interest rate plus a few per cent. Today this is about 20% — the figure we agreed that was needed in Chapter 3.

# Can we charge different prices for the same thing?

*Not all customers will receive the same benefit from our product*

Customers will not all perceive the same value in our offering, partly because they will not receive the same actual benefits and partly because they are receptive in varying degrees to our communication.

In the case of OPERATOR, the team could have started by realising that large and small foundries would have different benefit levels from the new flame control system, NewFound II. In addition, there would be a difference depending on whether simple or complex foundry work was involved; the former has much less wastage.

Again, a foundry that had already invested heavily in new furnaces might not be so receptive to OPERATOR's arguments.

OPERATOR might also find that the benefit of its new system would depend on what the customer was using at present. The new linear systems might have to compete with manual ones.

## Do all customers have to pay the same price?

> Depending on whether a customer has OPERATOR's linear system, one from a competitor or a manual system, it could be shown that the customer value would be £0.5, £0.6 and £1.0 million per year, respectively.
>
> Would not it therefore be foolish to base the price level on the lowest alternative, £0.5 million, as we did in the previous chapter?

Just imagine the results if we could somehow achieve three different price levels instead. The value of the market potential of customers with manual systems would double, but are we 'allowed' to price like that?

Can you really charge three different prices for the same thing?

## What we can learn from the airlines

I travel between London and Copenhagen frequently and I always have three different price levels from which to choose; around two hundred, three hundred and four hundred pounds.

Do you know the rules for APEX fares?

If there was the equivalent of a Nobel Prize for 'Pricing According to Customer-perceived Value', in my opinion it should be awarded to whoever invented APEX fares.

Imagine you were the marketing manager for an airline before the APEX fare was invented, and your boss gave you the following task:

'We have 25% of our seats empty on every flight to Copenhagen. Couldn't you sell these at half-price to travellers who would like to fly with us but cannot afford the full fare? But make sure that those that can afford to — mainly our business travellers — keep paying the full amount!'

As you may well know the answer you may think the task was simple, but that is with the considerable benefit of hindsight.

What is it that makes it possible to have different prices for virtually the same thing? There is no difference in the seats offered to full- and

APEX fare-paying passengers, and even business class isn't all that different.

Your starting point is of course the major difference in price sensitivity between those whose company is paying for their ticket and those who pay for their own. How can we differentiate them?

If we say that an APEX ticket has to be bought and paid for at least two weeks prior to departure, then we further stipulate that no changes are allowed and, as a final restriction, that you have to stay away over a Saturday night, we have created three effective filters.

Very few business people know all the details of their trips more than two weeks in advance. Even if some of them do, and get through the first filter, they normally get caught in the second; frustrated secretaries will tell you that every major meeting is rescheduled at least twice!

If someone unexpectedly manages to pass those two hurdles, there is the third one — staying away over Saturday night. Can you imagine the businessman or woman calling his or her spouse at home to say: 'Darling, to save some money for the company, I'm going to stay here in Copenhagen over Saturday night', thus spoiling their weekend?

## *The art of price differentiation*

Unfortunately, business life is not quite so simple as to allow us to copy the airlines directly. But we should adopt their philosophy of finding optimal product offerings for each customer or customer group.

In many businesses the comparison with an airline is very pertinent; with high fixed overheads you have to pay for today's production slots even if they are empty.

Each business has to find its own solutions on how to structure its offerings. Price differentiation is an art requiring both logic and creativity. The logic tells us that customers have different needs and therefore will derive different benefits from our offer. Put in other words, they have different degrees of price sensitivity. The challenge now is to create a pricing system that allows us to split the market into two or more segments. We then need to fix prices for each segment such that we persuade our customers to re-invest as large a part of the customer benefit as possible into continued development with us.

## Price differentiation in OPERATOR

The way in which we charge for our products or services has a significant impact on the possibilities open to us to differentiate the price.

Charging for solvent or consumables or the sharing of profit and risk automatically entails a price differentiation between large and small customers. These methods do not enable us, however, to differentiate between customers on the basis of the type of hardware they are presently using, which is probably the most important factor in OPERATOR's situation.

If you ask for all the money up-front, price differentiation with respect to previous hardware installations is simple. We take back the old system at an appropriate price, which naturally bears no relation to the condition or usefulness of the hardware, but which has been set at a level to reflect the customer's price sensitivity.

Not even a 'buy-back' offer scheme like this is a new price differentiation method. As well as being common currently in markets such as domestic appliances (cookers, refrigerators etc.), some time ago this method was used with electric razors; but did anyone really think that someone else would use the old razor they had traded in recently?

---

What should OPERATOR do now then?

It all depends upon who is the customer. In one specific case, Marketing Manager Leonard Hewson came up with the following solution for a customer with only a manual flame control system.

---

..... so therefore we offer you a complete system consisting of:

**THE COMPUTERISED FLAMENCATOR SYSTEM 'FLAME II'**
*including  configuration, installation, testing and training*
Investment: £980,000

**SOFTWARE LICENCE**
Monthly fee: £6900

**SERVICE AGREEMENT**
*including maintenance, updating and performance reports*
Monthly fee: £7200

**CONTROL SOLVENT 'FLAME CONTROLLER II'**
Price per litre: £0.62

Please see the enclosed pay-back calculation!

Leonard never misses a chance to show the customers how they can 'cash in on the investment'. If he catches one of his sales engineers leaving the customer to do the pay-back calculation, he feels so strongly that he nearly considers it to be a case for disciplinary action. His pay-back calculation runs as follows:

| | |
|---|---:|
| INVESTMENT | **£980,000** |
| INCREASED CASHFLOW/year | £885,000 |
| (decreased energy cost and wastage only) | |
| (NB. We have not assumed higher revenue | |
|     on account of higher quality) | |
| less licence and service costs | −£168,000 |
| less consumption of control solvent | −£140,000 |
| NET INCOME SURPLUS per year | **+£577,000** |
| | |
| PAY-BACK PERIOD: INVESTMENT/ INCOME SURPLUS = 980/577 | |
| | **= <2 YEARS** |

The recipient of this offer has the maximum benefit of OPERATOR's new Flamencator system, as the customer currently only has a manual control system. However, by including configuration, installation, testing and training in the offer, Leonard has created scope for price differentiation.

If the next customer already has our old system, an upgrade can be offered at a price that matches the increased income surplus. If the customer has a competing system, Leonard can offer a 'buy-back' scheme or a combination of some additional hardware plus adjustments in the software package.

# Your comments

*How does your company use price differentiation? Can you see any possibilities for making additional profits?*

# Delivery and payment terms

Strictly speaking, we have not defined a price fully until you know where the handing over (transfer of ownership of the product) takes place and when the invoice has to be paid. A price of 100 can be expensive if I have to pay in advance and be responsible for freight and risks in transit from San Francisco. But if I have a 90 day credit period and the entire system installed and ready in London, 110 could be a bargain.

## *Who should pay the freight?*

Making the customer pay the freight is a negative form of price differentiation. The customer is penalised for not being next door to us! We cannot very well claim that the customer's benefit from our products increases with distance from us.

The problem of how freight costs and risks should be divided has existed as long as there has been international trade. There are, as you may know, international conventions as to how these are expressed. Terms like ex-works, FOB, CIF and so on are becoming standardised under the heading Incoterms issued by the International Chamber of Commerce in Paris.

I am not going to discuss them in detail as I am sure there are people in your company who know more about them than I do. Rather my ambition here is some strategic discussion regarding delivery terms within post-industrial business development. Should we offer our systems ex-works, meaning that the customer has to take care of all problems beyond our doorstep, or should we offer DDP (Delivered Duty Paid)?

For me, it is clear-cut. To ask the customer to take care of all the problems and come to us to get their system 'ex-works' is not what attention to the customer is all about in post-industrial business development. We should take care of all our customers' problems, but only so far as is practical.

Stating it this strongly may be misleading because I have to admit that there can always be exceptions. For example, if the customer considers him- or herself to be better in international transportation than we are, he or she should ask for ex-works delivery.

Another reason to ask the customer to take on more (of what we ideally should do) could be that we simply do not care — if the particular market sector is not one we have prioritised, and if it is too difficult for us to sort out all the local rules and conditions. But then the question arises as to why we are there in the first place. This has been covered in Chapter 4 ('Follow the narrow road').

## *International payment terms*

When we discuss exports, certain special payment terms can become applicable, such as letters of credit. There is a multitude of special literature devoted to the subject.

In a post-industrial business environment, however, there is less emphasis on special export conditions as we should consider every chosen market as a home market, as will be developed further in Chapter 10.

---

### **Your comments**

*What delivery and payment terms does your company use? Do you see any reason to change them?*

# Currency considerations

*Can we drop currency problems in our customer's lap?*

It cannot be the customer's fault or problem that not all the world trades in Pounds Sterling.

In post-industrial business development, we should aim to solve all the customer's problems, including problems of currency. Naturally we should quote in the importing country's currency whenever possible and practical.

## The advantage of taking on currency problems

There are many advantages in taking on the currency problems ourselves.

In post-industrial business development, we see entry into a new geographical market as a long-term investment decision. Are we then going to let currency fluctuations dictate our pricing in that market? That is what happens if we only quote in Pounds Sterling.

Both distributors and subsidiaries tend automatically to translate prices according to the exchange rate. Our local representative seldom has the long-term vision to allow him or her to absorb any currency fluctuations within his or her own margin.

If we quote in local currency and the exchange rate goes down, we have to examine the competitive situation and decide whether to ride it out and wait until the exchange rate goes up again, or whether to increase prices. Our decision depends on the competitive situation and our ambitions of market share.

If our main competitors also produce most of the added value outside the market, they too may be expected to increase prices, whereas if our competitors are local we have to be more careful.

When currency fluctuations go our way there is absolutely no reason to miss the chance of improving our margins. If we want to maximise our marketing effort, we should use these improved margins on a more effective competitive weapon than price.

# Summary

In post-industrial business development, we are not the victims of price development — **we** develop the prices.

How much we get paid depends partly on how we charge and partly on how good we are at explaining why we should be paid this much.

If we want all the money up-front, we make life hard for ourselves. The customers' decision will, more than likely, have to be taken at board level and, as we supply large amounts of invisible values, we may have difficulty in explaining what the customers get for all their money.

It might be easier to be paid per month or per unit produced by the customer, or possibly per consumable item utilised, if these are appropriate to the application.

If we manage to convince the customer of the sense in re-investing a significant part of the customer benefit we create in continued development with us, we may find certain years showing higher than average profits. Such profits should not be considered excessive, but should be reserved to counteract leaner years when we next see increase in our development costs.

Not all customers will have equal possibilities to benefit from our offering and therefore we have to differentiate our prices. There are many traditional price differentiation systems such as quantity discounts, seasonal discounts etc. We cannot always use these, but have to find our own methods. We can learn a lot from how the airlines managed to solve the impossible problem of differentiating ticket prices for the same trip, using APEX fare restrictions etc.

When we use price according to customer-perceived value, we cannot drop delivery or currency problems in the customer's lap. They will hardly perceive a higher value simply because they happen to be located far away from us. Even if the customer does not explicitly ask for them, we should offer ambitious delivery terms, making it as convenient as possible for a foreign customer to do business with us.

In most cases we should quote our prices in the currency of the importing country, as we do not want currency fluctuations to govern our long-term pricing policy in any of our markets.

# Chapter 10

# Distribution development

*In the industrial society, extensive export was possible. Companies aimed to make money from small market shares in a multitude of markets.*

*In the post-industrial or information society, our goal must be to treat every chosen market as a home market.*

The traditional industrial product's route from producer to consumer has been described in many books. I imagine you are fairly well acquainted with the process.

It used to be possible to export extensively. By this I mean to sell to whoever wants to buy; sometimes on the condition that the customer will pick up the delivery at our factory, paying in pounds, preferably in advance. This could sometimes result in fairly small market shares in many markets, but these could together make up enough volume for efficient mass production in one place.

When marketing simple products, with low degrees of customisation, it is possible to produce the bulk of the added value in one place.

When marketing standard products, that are well-known in the marketplace, possibly only needing an instruction manual, the price is generally the most important competitive weapon. When this is the case, it is possible to calculate a price for every step in the distribution chain and internal prices between parent company and subsidiary work well.

To market knowledge-intensive components and systems in overseas markets creates new problems and challenges for Business Engineers. How we deal with these is the subject of this chapter.

# Choice of distribution route

## *Direct sales*

To sell directly to an end user in a foreign country can only be recommended for reference installations. By this I mean that a few key sales should be made this way, if necessary at lower margins, in order to get a foot in the door. Once we have a few local reference sites, our credibility will increase. Then we can consider using local agents for business at better margins.

Although medium or large sized customers do not experience problems in importing components directly, it can be difficult to serve a system customer properly without a local representative.

## *Agents on commission*

A commissioned agent of the traditional sort, who finds a customer and then helps with a few translations, is no solution for a company selling advanced systems, other than in the initial phases.

If the commission agent is asked to do much more work, his or her commission fee becomes quite high — often between ten and twenty per cent and then, I'm afraid, there can be problems.

Sometimes we will still want to work with people on commission anyway — the alternative is to employ them and we may not be ready to do this and start our own subsidiary yet. Or maybe we just cannot afford it.

Commission agreements can be an expensive solution and very difficult to terminate. I strongly discourage loose agreements of the type:

'Why don't you start working on commission until we've tested the market; we can draw up an agreement later.'

If you sign an agreement like this in France, you may find that legally you are stuck with it for years.

In many countries legislation exists governing incomplete agreements. This legislation comes into effect when the terms and conditions for cooperation between the new agent and yourself have not been spelt out clearly. In most cases these rules are to the advantage of the local representative. In addition, it is generally good business practice to know what kind of agreement you have signed!

If you decide to use agents on commission, you should be aware that you can never entirely trust what they say about your ability to charge higher prices (i.e. price sensitivity).

Let us assume that we are in the midst of tough negotiations with the purchasing manager of a new Dutch customer. We have offered a price of 500,000 Dutch Guilders. The purchasing manager says he can buy exactly the same product in Holland for 450,000. But because he's been to school with our agent, the purchasing manager will give us the chance to come in at the same price as the local competitor, rather than having to underbid them.

Under the pretext of having to look over our calculations, we retire and ask our agent, who works for us on a 4% commission, if the purchasing manager is serious, or whether we should hold out for 500,000, which already constitutes a 5% discount.

Our agent tells us that it is only through his good personal contact that we are being given this opportunity to come in as a supplier at the same price as the opposition. He advises us not to risk it, but take the chance now or risk being thrown out before the cooperation has even started.

How should we judge his advice?

In principle, I don't think that we should question our agent's loyalty to our company in any way. If we gamble on sticking to the higher price and risking the whole deal, it is a question of our business selection.

The problem is that the agent has quite a different starting point from us. In this case, when the customer says that we are more than 10% too high in price, the agent only has 11% to gain from the risk of holding out for the higher price whereas he has 100% to lose if we do not get the order.

Let's say our unit cost is 375,000; this means our contribution would be 125,000 at a price of 500,000. If we are negotiated down to 450,000 the contribution is reduced to 75,000, a reduction of 40%. We, therefore, have much more to gain (67%) than the agent from holding out for the higher price.

In a case like this, we may have to assume that the agent, consciously or unconsciously, looks after his own interests.

Does this mean that we only have limited use of our local representatives for establishing market prices? Yes, unfortunately, it does, which is disappointing considering that they may very well be the best people to tell us about local conditions.

We can, however, place the agent in a similar position to ourselves by changing the commission system. One radical solution is to give the agent commission on our contribution. This does tend to have the disadvantage of leading to long, drawn-out discussions of our internal cost structure.

Seldom used but very logical are so-called limit prices. These are prices the agents can exceed but not go below. The agents could, for example, have a 4% commission on the limit price and 25% on the difference between the limit price and any higher price they can achieve. If this system had been in place in the example above and the limit price had been 450,000, our agent would have had just as much to lose as we had in reducing the price from 500,000 to 450,000.

A limit price has many advantages:

1. As a company we have the opportunity to earn more;

2. The agent has an opportunity to earn more;

3. The agent gets extra commission only on that part of the price we have not been able to obtain ourselves;

4. The limit prices can be revised at least once a year so as to prevent the agent from having an indefinite bonus from the discovery that it was possible to charge a higher price;

5. We can fix a limit price at such a level that we give the agent an extra incentive to work hard with products that we want to activate, for one reason or another.

The one disadvantage with the system is a certain amount of additional administrative work. One could argue that the agent will always try to reduce the limit price to achieve extra commission, but this is no worse than in a traditional commission arragement where the agent simply wants to make his or her life easier.

When using limit prices in this particular example, we would get 75% of the possible price increase. On normal commission, this price increase would not take place so we would lose 100% of the extra it would have been possible to charge.

The limit method should be considered as early as at the market planning stage, when drawing up price-lists.

There is another method to put the agents in the same situation as ourselves. We can demand that they take the same cut in contribution as we do in any price reduction. The business will not become any more profitable for us, but this may motivate the agent to keep the prices at a higher level.

## Distributors and joint ventures

Once upon a time there was a company marketing simple, high quality, forged tools in the North American market. They sold through a distributor who bought at fixed prices.

All was well until the company developed a new system, requiring a minor modification of existing customers' equipment. Some time after the launch of this new system, the distributor asked for a substantial discount, otherwise he would not be able to afford to continue, he said.

The company could not understand his position. They sold to him at cost and knew that the distributor marked up the price by 100%. In addition, the modification costs were not included since they were offered separately. In this situation, the manufacturer questioned whether the cost of moving the machine system across the Atlantic to the end user in the United States could be as high as cost of manufacturing in the first place (and therefore merit a 100% mark-up).

The Americans had a very open attitude to the problem. They invited their partners to follow the sales process and see all the calculations and costs. It was apparent that the work spent in persuading an American end user to convert to the new system was both lengthy and very costly. But once they had been successful, selling was just a matter of ringing the customer once a month and asking how many tool parts they needed.

The manufacturing company had asked the American distributor to make a very large investment in the market without any guarantees.

The distributor was faced with the normal dilemma. If they sold too little they would lose distribution rights and if they sold too much there was a risk that the company would take over and start their own sales company, creaming off the sales from converted customers.

The situation ended up with both companies agreeing to start a joint venture, setting up a company where they each owned 50% of the shares. This, unfortunately, ended in disaster.

The idea was for the distributor to sell administration and sales capacity to the new company and for the manufacturer to finance the operation through a ninety day credit agreement.

The problem with the guarantees to the distributor had been solved but not the financing problem. The distributor was initially very pleased; he could invoice the new company for everything he did. However, the manufacturing company soon found that they did not get any money. This was simply because their new company could not pay in ninety days. The outstanding debt from the new company just grew and grew. The new company didn't earn enough money to be able to finance the growth with money generated by the North American sales.

Both parties had underestimated the size of the market investments, a mistake that is all too common. The early arrival of a recession, rather unexpectedly, meant that customers were postponing the conversion decisions — the sales development was not what had been expected.

The manufacturing company complained that the new company paid the invoices from the Americans but not from the British. The Americans countered by saying they didn't quite agree on the internal prices from Britain.

The whole venture ended with the Americans selling out to the British.

## An owned subsidiary

We now come to the alternative of an owned subsidiary — our preferred choice in a post-industrial business.

When selling sophisticated systems, entering a new market entails substantial investment. It is not always easy to accomplish this when working together with other independent companies.

And if the most important part of our product is software and service, it can be difficult to let someone else do it for us. However, it is not always possible for us to have exactly what we want.

An owned subsidiary is also a substantial investment, more than most companies normally imagine. There are numerous examples showing how companies have been forced to give up halfway, with serious waste of capital resulting.

If you do not know that you have enough money to take you all the way, you may have to be satisfied with the role of component supplier to an independent distributor who will supply the added value that comes from customisation to the end user.

Another alternative is to delay the launch into the new market until you can afford to do it well yourself.

So! Why not stick to the straight and narrow and avoid branching out into too many business units?

## Your comments

*What distribution strategy does your company apply? Do you think it could be improved?*

# The sub-optimising subsidiary

Let's go back a couple of years in time, when OPERATOR was setting up in North America.

Leonard Hewson was happy. He had what anyone would love to have in a post-industrial business; his very own subsidiary in Boston.

OPERATOR had appointed a British MD (or President as it is called over there) — Mr. Leonard Hewson himself. But in most other respects the company tried to behave 'as American as possible'.

It was first suggested that the company would be called OPERATOR NORTH AMERICA Inc. as this appealed to Charles Armstrong's imperialist nature. Fortunately Leonard managed to convince him that it would be detrimental to business to state so bluntly that the owners were foreign, so it ended up being called OPERATOR Inc. instead.

From the outset it was clearly stated that the British MD was there only to start up the activity and that he would hand over to someone else within three years, most probably an American.

## *The danger of doing business with yourself*

The whole investment in the North American subsidiary nearly went down the pan because of two misunderstandings.

First, Charles Armstrong thought that you have to put people under pressure in order to get them to do their best; using either the stick or the carrot. Second, he thought that it was possible to split the profits 'fairly' between the parent company in Britain and the subsidiary in North America.

Charles realised that the stick couldn't reach Leonard in Boston, so he tried the carrot. Leonard's salary would to a large extent be dependent on OPERATOR Inc.'s results; he would be paid based on NBT (net profit before tax).

This naturally invited difficulties and conflicts. Leonard Hewson, being an intelligent man, soon realised that the internal prices set by the parent company would to a large extent decide his salary. Too much of his time was therefore spent negotiating with the parent company, rather than with customers. His whole behaviour pattern became distorted because his picture of the company's income and cost structure had also become distorted.

Consider the cost structures as seen from the two viewpoints:

| **Parent's consolidated view** | **Subsidiary's sub-optimised view** |
|---|---|
| Income: 100 | Income: 100 |

| | |
|---|---|
| Contribution: 70 | Contribution: 30 |
| Variable unit cost: 30 | Internal price: 70 |

To ensure that he had sufficient capacity, Leonard presented a high-volume budget at very low margins, calculated just to make an approved bottom line figure.

In reality, he then adopted a policy of high price and limited marketing investments. He had neither the margins nor the balance sheet for anything other than that.

When the first annual report came, it was obvious that Leonard had totally missed his volume budget.

When Charles took him to task over this, Leonard could show, however, that the high prices and low market investments had given him a good NBT figure.

'And you did say it's the bottom line that counts,' Leonard grinned broadly as he spoke to a furious Charles.

Charles Armstrong had every reason to be dissatisfied but he was himself partly to blame, for he had written the rules.

When Leonard went for NBT rather than volume, the arithmetic went his way since a large part of the costs had been classified as unit costs and hence variable. The lost volume, however, reduced the parent company's capacity utilisation since the company's proportion of unit costs in reality only came to about 30%.

What was even worse was that OPERATOR couldn't reach the market position it had aimed for and had therefore lost ground to its competitors.

What should Charles have done in this situation?

First of all I think it is time to realise that people like Leonard very seldom under-achieve. More often, it is the other way around. The desire to make a brilliant individual contribution is sometimes so strong it will affect teamwork.

Second, we should realise that people under stress often do not perform well. It would have been much better if Charles had shown an attitude of trust towards Leonard, and let him feel the appreciation he deserved.

But what about the money?

In this particular case, I think Charles should have given Leonard some compensation for the difficulties of moving to the USA for a while. And Leonard should be paid on his personal performance rather than on the performance of the subsidiary as those results are too strongly influenced by many factors outside Leonard's control.

## Don't take internal prices too seriously

When we sold simple, tangible products, it was easy to assign a realistic result to a subsidiary, possibly comparing it with the alternative of selling through an independent distributor.

The more sophisticated our products and deliveries become, the more aggressively we have to invest in the market in order to establish our position. At the same time it becomes less relevant to try to judge our subsidiary's result purely from the annual accounts.

Internal prices are set more to fulfil the requirements of the host nation than to give a true picture of local profitability. The fact that the figures can easily be 'adjusted' makes internal pricing interesting only to those in charge of co-ordinating a company's local taxation.

## The break-up of the subsidiary

We have seen in Chapter 4 that we should aim to become world leaders in a special niche market.

Unfortunately, we seldom succeed. We have great difficulty in concentrating sufficiently; we want to get into lots of interesting markets.

In order to defend our lack of concentration, we sometimes say 'it's good to have more than one leg to stand on'. I think it is difficult to justify more than two legs but some people think that having four legs is ideal.

When the number of special markets is more than one, the subsidiary will break up, as the 'knowledge' dimensions of technology and applications (see the Ansoff diagram in Chapter 4) are dominant in post-industrial business.

Back in our example, so long as Leonard was in the USA, OPERATOR Ltd. had no problems there as the business was almost exclusively flame control systems for energy savings in foundries.

But what should happen when there are two business divisions — remember OPERATOR has both METAL and PULP divisions?

A division should as far as possible have independent resources, but OPERATOR could definitely not afford two subsidiaries in the USA.

## *Think globally, act locally!*

So we find ourselves in the inevitable matrix organisation. For OPERATOR it looks something like this:

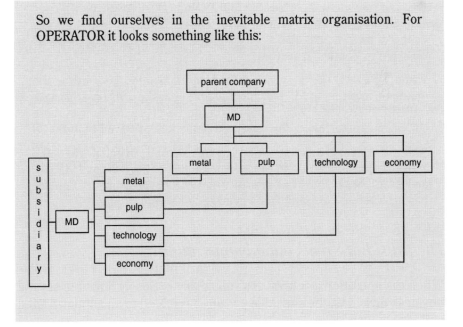

Commenting on the matrix organisation, Percy Barnevik of ABB wrote the following in *Harvard Business Review*.

> 'The matrix is the framework through which we organise our activities. It allows us to optimise our businesses globally and maximise performance in every country in which we operate. Some people resist it — they say the matrix is too rigid, too simplistic, but what choice do you have? To say you don't like a matrix is like saying you don't like factories, or you don't like breathing, and, as with these things, the matrix is a fact of life.
>
> 'If you deny the formal matrix, you wind up with an informal one and that's much harder to reckon with. As we learn to master the matrix we get a truly multi-domestic organisation.'

The people who manage the business in the subsidiaries therefore have to live with two bosses — the local manager of the subsidiary and the global divisional manager.

## Divisional managers are in control

Leonard Hewson was replaced by Bill Thompson, an American who was a foundry specialist. He was hired by Leonard, and Bill therefore saw Leonard as his real boss even though OPERATOR Inc. formally reported to Charles Armstrong.

When OPERATOR was divisionalised and appointed Leonard as the divisional manager of METAL and Ken Parker as the divisional manager of PULP, there were some problems in the US office. Ken's man there complained bitterly over the cavalier treatment he received from Bill.

After a stormy meeting, Charles Armstrong decided that the subsidiaries should be split as far as possible into the divisions METAL and PULP, with the local manager for PULP reporting to Ken for business and to the subsidiary's manager for administration.

## The subsidiary rents out administration?

When the operational activity of a subsidiary is closely linked to several business divisions, the subsidiary is often reduced to an administration unit, as we have seen.

One sensible way to look at it is to regard the head of the subsidiary as renting out administration capacity and upholding local rules and regulations.

## The skilled and much-needed office manager

In the case of OPERATOR, Charles Armstrong's forceful intervention solved the situation but, in many larger companies, divisionalisation of subsidiaries meets with considerable problems. In particular, when the divisions are large and comparable in strength, it is often difficult to move one of the divisional managers to head the subsidiary.

To be appointed 'office manager', without the power to make strategic business decisions, is seldom viewed as an attractive post by those who have previously held decision-making responsibility.

The desire to give orders is still strong in many people, even though lip service is given to the principle of management by objectives. We still have not been able to tear down the pyramids! Sitting at the top and controlling everyone else still seems to be the most attractive objective to many people.

The more our company develops towards a post-industrial or knowledge company, the harder it becomes to uphold this old style of leadership. What we need now are team leaders and mentors.

The role as local head of a subsidiary does not have to be a janitor's role, just serving divisional managers who report straight to Head Office. Often divisional managers are specialists in their own field and they may be from another country. They need the strong support of a local, knowledgeable and skilled person who can sometimes function as a mentor.

The head of a subsidiary, in addition, has the important responsibility for continuity. He or she often has sufficient qualifications to hold the post of divisional manager temporarily, should the present incumbent disappear or move.

In most companies, there is a greater turnover of personnel in the more specialist positions than in administrative ones. Segmentation of the market is not static; as companies grow and contract, this gives rise to frequent movements of key personnel.

In short, to head a subsidiary is a very skilled role, even in a divisionalised organisation.

# Licensing and franchising

In the transition from the industrial society to post-industrial or information society, the ratio of information content to material mass tends to infinity. I have already mentioned as an example the transition from punch-card computing machines to modern PCs and lap-tops.

The extreme examples are software and consultancy services. The value per kilogram of consultancy service must be infinite unless the weight of the consultants themselves is included.

You may be asking what point it is that I am driving at here. The point is that the more sophisticated the systems become, the more apparent it is that in reality we are talking about an international trade in knowledge. If it wasn't for the fact that it is very difficult to get paid for pure knowledge, we would need to send even less material across the world.

So why not sell the knowledge separately?

Do you remember the European company that developed the system for particle measurement that I cited in Chapter 8 — the American with the salad bowl and the sugar? It was very successful and, having sold a few measurement installations in Europe, it was time to deliver to the US too.

The basic idea was a new method of image processing, but to make the system work, cameras, light sources, computers, tables etc. were all needed. It turned out to be very difficult to send the full set-up from Europe to the US; even the mains voltage and frequency are different. In addition, excellent components existed in the US which had been developed for American conditions.

## *Licensing*

'Why not sell the solution for a licence fee all over the world?' someone asked, eventually.

Licensing is a tried and tested method of raising revenue from an invention without having to get involved in the nitty-gritty of local business. Of course you can go down that route, if you are satisfied with a small part of the value of the solution. Unfortunately, those taking out licences tend to have a mental block against two digit percentages when it comes to fees paid. Often the inventor or originator is only offered one or a few per cent.

But a few per cent of what?

An effective way of cheating the inventor is for the licensee to limit the scope of the deliveries on which licence fees are to be paid; for example, selling just the licensor's printed circuit boards and cameras and selling the ancillary tables, cabling, installation, training, service etc. from other sources.

The disadvantage of licensing in the post-industrial economy is that the base for our licence fees is set to become smaller and smaller as the proportion of 'hardware' goes down.

This does not mean that licensing cannot be useful — sometimes there is really no other choice from a financial point-of-view.

## *Franchising*

In a rather simplistic way, you could say that franchising is a wider and more comprehensive form of licensing. This is where the franchiser complements the licence with an additional package comprising training, use of a brand name, marketing support etc., often put together under a registered trademark.

Whereas licensing virtually presupposes a patent, as the technology is being sold almost raw, much more than this is sold in franchising. Through the extensive documentation, software content and protection afforded using a trademark, franchising will not be as dependent on a patent as licensing.

If our friends with the measurement system had decided to use franchising, ideally they would have registered a trademark and built up a substantial 'software' package. Their specifications could decide the scope of the franchisee's deliveries of tables, installations, hardware etc., to minimise argument over what exactly the licence fee would be based upon.

Franchising as we know it today started in earnest at the beginning of the 1970s. In recent years it has grown strongly in many areas of business and it can be an excellent way to develop 'knowledge' business. This will particularly be the case where products (deliveries) require substantial implementation and customisation activity carried out by highly skilled and motivated individuals close to the end-user.

# Summary

Knowledge-intensive products and systems are difficult to 'ex-port' — in the traditional meaning of the word, i.e. shipping something out as a standalone finished product.

Today, new markets require a very different involvement from any company wanting to get established. The complexity of systems and products and the requirement for customisation means that a substantial part of the added value has to be achieved locally. At the same time, the marketing has to be global to achieve a large enough base to build up sufficient revenue for effective system development.

'Think globally, act locally' is a recommendation often heard in this and other contexts, here translated as the need for 'multi-domestic' business development. Knowledge-intensive products and systems have to be positioned in their markets. You cannot achieve a strong position if you spread your activities too thinly on the ground with only limited marketing investments in each country.

The best strategy of all would be to have your own subsidiary in each country, which means major investment. How the profitability of this investment should be calculated is no longer obvious as traditional internal price structures hardly show anything useful other than how local taxation can be optimised.

Whichever way you calculate, most companies find that a newly established subsidiary is more expensive than anticipated. If you are not certain of being able to persevere with it, you may have to rely on a distributor to sell your system components instead. But if the distributor is forced to invest substantial sums in marketing, you may find yourself back with the original problem.

A better alternative could be to wait until you have made enough money in your other, earlier, markets to finance a determined and well-funded entry into a new market.

Franchising is an excellent way to develop 'knowledge' or post-industrial business, particularly if your deliveries require substantial execution and customisation work by highly motivated and skilled individuals located close to the end customer.

Franchising can be considered to be an extended form of licensing where the franchiser complements the licence with a 'software' package comprising training, use of a trademark and/or brand name, marketing support etc.

Chapter 11

# Communication development

*The post-industrial or information society is considered by many to be synonymous with the communication society.*

*It is therefore natural to devote a chapter to discussion of the post-industrial company's communications — not just with its customers but with all the stakeholders and other parties who influence the company.*

The word communication means different things to different people. It can mean roads and airlines or it can mean computers, databases and telecommunications. When I discuss the development of a company's communications, I am thinking about the exchange of ideas between the company and its interested parties.

It is impossible to exaggerate the importance of efficient communications in the post-industrial society.

However, I will restrict our treatment here to communication in its more abstract sense; my aim in this chapter is to give a post-industrial perspective on how a company should communicate with its interested parties.

# Communication with our employees

In an earlier chapter, when we ranked the interested parties in a post-industrial company, we started with the customers. This time we are going to start with the employees.

The post-industrial company can be characterised by the large invisible content of its deliveries to customers. If our employees do not know what we are trying to deliver, our customers certainly won't.

## *The importance of understanding*

As we saw in Chapter 8, communication with and between our employees should start right at the earliest stages of development of new systems. The effectiveness of our system development is strongly linked to our ability to communicate.

A post-industrial company, particularly, is best run through 'management by objectives'. The organisation's goal orientation and effectiveness will therefore be a function of the management's ability to communicate what should be achieved.

## *The importance of acceptance*

It is not sufficient simply that we make our people understand. Many individuals today do not just follow instructions; they have to be convinced and accept guidelines from management in order for them to make things happen.

When it is no longer a matter of easily quantifiable routine tasks, but work that requires creativity and active participation, strategy and guidelines have to be sold to staff in order for them to give their best.

One fruitful way of developing an organisation is to define both internal and external customers.

## *The importance of two-way communication*

If the communication from the company to its staff is to be accepted, the company must also accept communication from the staff.

It would be hubris by the management, however, to consider communication from the staff just as the price extracted for acceptance of their own communication. It is, after all, only through the knowledge, experience and views of the staff that the company develops at all.

Therefore the company has to learn to listen, actively.

## *The importance of having time to talk to each other*

Company magazines and suggestion boxes are good in their way, but we must also have time for internal personal communication — an atmosphere where the pressure of work does not allow for any casual contact can easily lead to individual alienation.

Certain creative processes positively require time for unstructured conversations. In one development-intensive company that I know, it is considered worse to miss a coffee break than a scheduled meeting, although this is perhaps taking things to extremes.

## *Using the media: the importance of timing*

The proverbial old woman who knew it must be true 'because she read it in the paper' is well-known to be naive.

Today we can see many examples where large corporations use PR campaigns in the press and on radio and television to boost the morale of their own people. It is easy to underestimate the importance of the feeling of pride that employees can feel when they see their company's name and logo in the media.

However, care has to be taken to avoid communicating something to external customers via the media before it has been agreed and communicated internally. It will be wasted effort and may well be positively harmful.

## *The importance of a strong corporate culture*

The stronger the corporate culture, the more strongly the employees will feel they belong to the company and the more they may see themselves as important to its operation. This is very valuable, in both good times and bad.

---

**Your comments**

*What proportion of its communication or advertising budget does your company spend on its employees? Is there a reason to change this?*

# Communication with our customers

## *We are better at external communication than internal*

In today's competitive marketplace, it is unnecessary to stress the importance of good communication with customers. Most companies already spend large amounts on advertising and promotional and sales activities.

'I know that half of our advertising budget is money down the drain, but I don't know which half!' is a lament heard from many MDs.

The fact that most direct mail advertising gets thrown in the bin is nothing to be upset over.

## *From diffuse to targeted communication*

The market is becoming increasingly sophisticated and demanding, so it is vital that our marketing keeps pace. We have a growing demand for selective communication with defined target groups.

Certain media are better suited than others to selective marketing. We will often find ourselves compelled to use direct mail-shots, using mailing lists of one type or another. But even if we select critically from a list, we cannot be sure that we will find only the prospects who are worth pursuing. That is why we should be grateful if those who have no interest in our offering throw the advertisement in the bin — we only want answers from those who are worthy of pursuit.

## *From mass communication to personal communication*

In mass marketing, mass communication is by definition a central part of a company's external communication strategy.

When marketing advanced systems, however, the emphasis is more towards personal or customised communication. The more sophisticated and customised the solutions, the more decisive personal contacts between buyer and seller will become.

## *The role of the written quote*

When we sell systems, the written quotation increases in importance in our communication with customers. The more complex the purchase, the more people participate in the customer's decision-making process. It is therefore not enough to use the old type of quote which served only as a legal instrument, telling the customers how much they had to pay and which responsibilities they were accepting.

## *The death of a salesman*

> 'He's a man out there in the blue, riding on a smile and a shoeshine'
>
> Arthur Miller, 1949, *'Death of a Salesman'*

The sale of simple, standardised products, chiefly through price, can be achieved with nice manners and a winning smile. To do successful business with advanced components, services and systems, you need a lot more.

The post-industrial company needs BUSINESS ENGINEERS.

## *The Business Engineer — the company's new hero*

Would you not agree that the individual who can convince a customer to pay £2 million rather than £390,000 for a system has to be one of the most important people in the company? (These numbers, you may recollect, appeared in our discussion in Chapter 9.)

When all the key positions in a company have been filled with business engineers, the company will become invincible in the market.

I think it is no overstatement to claim that one of the most important tasks for top management is to develop as many of the employees as possible into Business Engineers and to give them the right environment in which to flourish.

How this can be done will be discussed in Chapter 13, 'Organisation development'.

## Communication after purchase

All too often, we abruptly cut off communication with the customers as soon as they have given us the order. This is particularly dangerous when selling systems where design and development work have been substantial.

As much of the value of our products is to an extent intangible, we have to make sure that our buyers use and understand the benefits and experience the value to the full. If our customers are dissatisfied, it is important to correct the situation immediately, as references will be crucial to future sales and thus our future business development.

---

## Your comments

*Do you still employ 'Terry, the travelling salesman'? Does your company have enough Business Engineers —- or would it benefit from more? Do you continue communication with a customer after the order?*

# Communication with our shareholders

*We cannot be sure of keeping our capital*

We cannot automatically assume that money will stay in our company just because we have earned it. It is the task of the capitalists to direct capital to where it can make best use of opportunities available. However, historic profitability is an important criterion for judging how to invest in the future. Any management team in a company which has had years of low profitability will have a credibility hurdle to overcome.

'Next year, we're sure it will improve...'

The more our company develops towards a knowledge or post-industrial company, the higher our fixed costs become. The life-cycles of our products or systems are unlikely to lengthen and turbulence in the marketplace is likely to remain commonplace. As a consequence, the problem of uneven development of profitability will probably become worse.

If, in addition, it is unlikely that every development project will lead to success, the picture is clear; in the future we have to accept that our results will be like a rollercoaster ride — at least using today's traditional accounting methods.

We will have problems keeping our capital both when we are at the top and at the bottom of our profitability curves. When we show red figures, it is likely that the capitalists, particularly suppliers and banks, will restrict credit to us. When we show a healthy profit, we may have to pay management fees to our parent company to help others less successful than ourselves.

Once we have run out of 'clever' accounting, the only defence against drain of our capital is good communication.

## The bank

A mistake we tend to make is only to communicate with the bank when we need to borrow money; that is, when we are least credible.

The bank should be kept informed of our plans and projects continuously. By showing foresight in our economic planning, we can anticipate our borrowing requirements well in advance and thereby greatly increase our possibilities for optimal financing.

## *The shareholders*

Different ownership scenarios can have dramatic effects on the possibilities open to a company to retain its capital. Compare the situation of a managing director in a family business with independent financing, to the position Charles Armstrong has in our example. Charles constantly has to convince the owners of the advantage of letting money stay in OPERATOR. In the next chapter, you will see how decisive this question can be for OPERATOR's strategy.

Shareholders are guaranteed information at least once a year in a company's annual report. When we observe the attention that is paid to the design, layout and production of annual reports, it is clear that many companies have understood the importance of good communication with their shareholders.

## *The board*

The most direct channel to the shareholders is often through the non-executive directors on the board, as they are appointed by the shareholders at the AGM.

The nature of the work carried out by a board has developed dramatically over the years. Board meetings used to be uneventful gatherings where some board members began the meeting by discreetly opening the envelope containing the agenda under the table. Nowadays boards in well-run companies often comprise a group of well-prepared individuals ready to support the managing director in all sorts of strategic questions.

A sensible MD is careful to involve the board in most or all strategic decisions and to 'sell' the opportunities for the future.

In companies without independent financing, it is of particular importance to be able to explain variations in profitability, for reasons we touched on earlier.

# Communication with our suppliers

Many are of the opinion that spending time and effort on communications with suppliers is not worthwhile; isn't it up to them to keep us happy as we are the customers?

I think this attitude is wrong. It is just as important to buy on favourable terms as it is to sell well. And favourable terms are usually achieved through good contact.

## Suppliers

Even though we may only have sporadic contacts with a supplier, there are good reasons to nurture contacts with the supplier's key people so that they understand our needs and prioritise us in all respects.

This is in particular true for new developments.

> As you may remember, OPERATOR's new semi-linear flame control systems were based on new developments from the supplier of sensors and processors. Had the supplier decided to go to one of OPERATOR's competitors first, the situation could have been radically different.

## Subcontractors

When our company is more firmly established in a network of subcontractors, the importance of good communication is even greater than for a simple supplier-purchaser relationship.

# Your comments

*How effective is your company's communication with its suppliers? Do you only talk to them when you need something?*

# Communication with society

Any company is a part of and plays a role in society and should not avoid communicating with it. To what degree and with whom to develop communication is, however, to a great extent dependent upon the company's activity and size.

## *Government*

Many companies have to communicate in some way with their government — even if most judge their size to be too small to do it directly. If this is the case, they have to do it through organisations within the industry such as trade associations or interest groups.

Government defines the framework within which the company can act and it is of the utmost importance to be able to influence this framework. As an example I would like to stress the importance of showing those in power the need for stronger balance sheets in the post-industrial economy.

## *Local authorities*

At local levels, even small companies have reason to keep direct contact with (local) government. Here the questions are often of a practical nature; the importance of good contact, for example when seeking planning permissions and licences, is obvious.

The 1990s are considered by many to be the decade when the greatest problem for companies will be to find skilled labour. Good contacts with academic institutions and the education and training sector could be crucial to the company's future.

## *The public*

It is likely that the only companies that can afford to communicate directly with the public will be those in the consumer goods sector. No company, however, can afford to give a bad image or reputation and most will benefit in some way from some communication of the positive values or aspects of the company to the wider public.

# A new role for the advertising agency

## To inform in the information society

Many companies use an advertising agency far too rarely, often approaching one far too late in the business process and generally only to communicate with customers.

If the post-industrial company is active within the 'information' society, what could be more natural than to use a professional communicator when needed?

## Professional communication is not only for customers

If you are ever forced to retreat strategically and dismiss staff, it is, as we saw in Chapter 7, very important to communicate the fact that it is only a retreat so that the remaining personnel do not lose faith in the company.

As we saw in Chapter 8 on system development, we need professional communication right from the start of evaluating a new system concept.

In Chapter 9 we found that when pricing according to customer-perceived value, how much we can charge is directly linked to how well we can communicate the advantages of our offer or product. The advertising agency therefore will have an important function in our pricing and revenue generation.

In Chapter 10, I related a story about a joint venture that failed. Perhaps even this could have been avoided if professional help had been sought to communicate better the calculation methods used in the internal pricing.

## Professional communication is not expensive

You often hear that it is expensive to use professional communicators, but these costs are low compared with the costs that can be incurred by bad communication.

When I suggest using an advertising agency, what I really mean is that one should use professional communicators. Whether this means an external agency or whether it is employees in the company's own advertising department is immaterial. The trend, however, is in the direction of leaving specialist activities like this to experts so that we can concentrate on our core business.

Either way the message is to use professionals when it is important to have effective communication.

# Summary

By definition an information or post-industrial company is trading within the information (or post-industrial) society. Its skill in communication may well decide its future.

Products themselves are largely invisible; in these days of micro-electronics the smaller seems to be the better. Services cannot be seen at all, whilst what the customers can see is often not what the company can or wants to sell.

Having realised that communication is so vital for the success of a modern company, isn't it strange that such a large part of it is often left to amateurs?

Many companies regularly elect to use professionals in an advertising agency to design advertisements aimed at customers, but few involve the agency in the product development and pricing stages of the business process.

We have agreed that the modern company should price according to customer-perceived value (CPV). The price then becomes a function of **functionality x communication**. If you only communicate part of the value... — do I need to say more?

But before the post-industrial company can communicate anything to its customers, its communication with its own staff members has to have been successful, as it is their acceptance of goals and means, and their creative participation, that are the basis for communicating value to customers.

As traditional book-keeping does not always show a true picture of how a company is doing, communication with capital owners has to work better to allow the company to retain or raise enough capital for long-term growth.

The advice to try to become 'world leader in a speciality niche area' is often repeated throughout this book. You cannot be the best in all areas — this means it is important to be good at buying from subcontractors and to form strategic alliances. Good terms and conditions stem from good relations; good relations are created by good communication.

The company is a part of society and must naturally communicate with it. In the long term, contacts with the academic world and the education and training sector can be crucial to find the skilled personnel upon whom post-industrial companies are so dependent.

# Chapter 12

# Strategy development

*In post-industrial business development we aim to become world leaders in a niche market. Frequent strategic assessments of market segmentation and resource utilisation therefore become important,but strategy development must never deteriorate into an academic 'playpen'.*

*In this chapter I present a synthesis of current strategy models to arrive at a practical tool for drawing up functioning business development plans in the shortest possible time.*

You can view this chapter as a practical summary of what we have discussed so far.

My dilemma is not knowing whether you the reader have already read ten books on strategy development or whether this is the first time that you have really looked at the subject systematically. Strategy development is, however, something that I am convinced you will have considered in real life, even if you used another name for it.

If you already are a 'strategist', I ask you to regard this chapter as my opinion of what is practical in post-industrial business development.

## Strategic planning — a waste of time?

> Charles Armstrong was very sceptical about strategy planning.
>
> 'We haven't got the time for those academic games!' he snorted bluntly when a new employee came up with the suggestion.

Charles Armstrong would have been around in the sixties when a few young, starry-eyed MBAs invaded many big companies, bringing with them their methods of extrapolating trends to produce strategic plans. In those days, MBAs were 'special' and had wisdom. Sometimes the plans they made were sound, but then along came the oil crisis of 1974; all the trends seemed to break down and more often than not the plans went wrong.

### 'Paralysis through analysis'

What the likes of Charles Armstrong resent most in long-term planning, as he would call it, is the paralysis it can generate inside the company.

The statement 'the question is being analysed by the strategy group' is often an excellent excuse for weak managers to avoid making difficult decisions.

### How to avoid paralysis

The way to avoid 'paralysis through analysis' is to see strategic planning as an ordinary business tool, without therefore trying to belittle or exaggerate its importance to the company.

It is inevitable that certain decisions will have greater importance than others and, if we do not take these decisions in proper fashion, others will do so for us — our competitors, for example.

In reality strategic decisions do not wait for the next strategy group meeting scheduled for May. On the contrary, they tend to creep in amongst daily decisions:

'Should we fill the vacancy left by Smith's departure, or would it be better to take the opportunity to reorganise?' is a typical example.

To answer this question correctly, there should be a strategic action programme developed through systematic strategy development.

After a 'strategy kick-off', where discussions start to mean more than just documentation, the process should become continuous and be updated continuously. In this way the strategy becomes a support and a common frame of reference, enabling key people to take the right strategic decisions without having to consult the group each time.

## Hypotheses

One way to avoid paralysis and save time is to work with hypotheses. All too often, you get caught up in analysing today's situation and describing where you stand; there are so many things you seem to need to consider before you can take a decision. My advice is to base the first round on hypotheses.

Unless management is new to the industry, there is probably a sound baseline of factual knowledge and 'gut feeling' within the organisation. Building on this, it should be feasible to complete the first round quickly and decide on which hypotheses are crucial to the strategy. Thereafter, more time and money can be spent on verifying these hypotheses.

## Shall we complicate it ourselves?

I have known several marketing managers relate the joke: 'Shall we hire a consultant or shall we complicate it ourselves?'

If you have nothing much to do and like complications it is a good idea to start discussions on strategy development without a consultant. The discussions can be endless and often result in 'refreshing' differences in opinion. However, the experience of an external consultant to structure and referee discussions will usually improve the effectiveness of the process.

Complete re-appraisal of strategy every three years, with intervening annual updates, is a common pattern. With only this amount of direct strategic planning experience, an internal team may not gain sufficient competence in strategy development. An outside member of the group can also minimise the risk of 'home blindness' — the difficulty of seeing routine or regular internal problems.

*Integration with the budgeting process*

Systematic strategy development can also save time in the budgeting process, incidentally.

If there has been no strategic planning *per se*, change and development needs to be considered in the budgeting process. However, many organisations simply insert new numbers in an old strategy to derive a new budget, thereby missing another prompt to plan strategically.

# OPERATOR Ltd. — an excellent example

In the remainder of this chapter I will describe how I think a strategic plan should be conducted. I emphasise the word 'think' because strategic planning can be carried out in many ways. I am sure that you have your own ideas but I hope you will find my comments helpful.

In my experience, time for good strategic planning is in short supply. Those who have enough time are staff functions (i.e. business development people rather than line officers such as sales directors etc.) and consultants, but they cannot themselves create the strategy. It is therefore important to concentrate on only the relevant issues and quickly get to the core strategic questions.

So as not to bore you with an abstract discussion of procedures, I have chosen to illustrate my ideas through the strategic plan of OPERATOR Ltd. as an example. As we shall be focusing on OPERATOR in this way, the typefaces used in the text will differ from previous chapters; where the text is the courier typeface this denotes extracts from OPERATOR's plan, whilst the normal typeface will be used to discuss implications for both OPERATOR and business in general.

To help you keep your bearings in the strategic plan and process, there will be an icon on the page indicating which step we are considering.

# Our economic starting position

STRATEGY
DEVELOPMENT

Economy ◄—

History

Present
situation

Business units

Competitive
analysis

Evaluation of
business units

Prioritising the
business units

Goal setting

Definition of
core questions

Execution

This section aims to establish our position in the business orientation matrix we introduced in Chapter 7.

Our (current) economic position is the starting point and basis for our strategy development. We should therefore do nothing until we know whether we are presently heading for bankruptcy or if we are in a position to attack our competitors' market positions aggressively.

## Liquidity

OPERATOR's liquidity is good according to the latest annual results. The figures, however, do not reflect the situation accurately. The worst month is August — if nothing is done about our credit and stock turnover ratio, we will need an increased overdraft before the holidays are over.

## Financial strength

The equity ratio in OPERATOR is 24%. An invisible asset is our property portfolio which is shown in the balance sheet at too low a value, but we must take into account our borrowing requirements to ensure sufficient liquidity in August.

Overall, we can conclude that our financial strength is too low. We need an average equity ratio of 30% to be safe; this means we should have around 40% before starting any aggressive action to increase market share.

## Profitability

Profitability on total capital — the return on capital employed (ROCE) — in the most recent annual accounts is 11%. It has fluctuated between 10% and 15% in recent years; enough to pay our interest bills but hardly enough to give us the opportunity to consolidate our financial strength. The dividends paid out to shareholders have always been very modest. It is unclear what INVESTOR plc will demand in the future.

It will be important to persuade INVESTOR to let us keep the profits we generate.

## Capacity utilisation

At the present moment, we could increase sales by up to a third without new investments in fixed assets or additional recruitment, were it not for undercapacity in the system programming area (Department S).

System programming here means the process of final customisation of a flame control system to the client's particular application.

## Bottlenecks in production

Due to the ever-increasing demands of the quality specialists, flame control systems are not considered to be ready for delivery without extensive final bespoke programming, which requires both time and skill.

Although overall we have overcapacity in the company, a backlog is piling up in Department S (S as in System Programming). The reason for insufficient capacity can be described either as too few people or too little computer support.

The CAD-CAM installation OPERATOR is using is outmoded, but upgrading it would be impossible. A completely new installation, costing between £0.5 and £1 million, would be necessary.

## Our economic starting point

We can conclude that we have the most difficult starting point for our strategy development: a red light and undercapacity. Even though customers are clamouring for our products, we aren't making any money.

One obvious strategic question will be how to increase capacity in Department S without having to make million pound investments.

STRATEGY
DEVELOPMENT
Economy ⟵
History
Present situation
Business units
Competitive analysis
Evaluation of business units
Prioritising the business units
Goal setting
Definition of core questions
Execution

Here it was no longer possible to stop Charles Armstrong from taking action before the strategy work was finished, and he should not, of course, be criticised for this.

'All our strategy development,' he began, 'is dependent on how successful we are in increasing capacity in Department S, without hiring more people or spending large amounts of money we don't have.'

Having made up his mind, Charles decided to 'lock up the engineers' in their laboratory, threatening to make them stay there until they had solved the problem.

Under this pressure the project group came up with a solution. They managed to increase capacity in Department S by 30% without hiring extra staff and with a computer investment of only half a million pounds!

Charles himself helped accomplish this productivity increase by transferring two of his brightest female engineers to Department S. The department had developed into something of a mutual admiration society, basking in the light of its importance to the company. The arrival of the new female influence changed the culture radically — to meet the new situation they simply had to shape up. Coffee breaks famous for high-flying theoretical system discussions were reduced and it was agreed to subcontract work that had previously been considered impossible work for anyone outside the department.

The section on the economic starting point for strategy development could therefore be rewritten as follows:

## **Our economic starting point (version 2)**

We can conclude that we have a red light and overcapacity.

What do we do in this situation? Do we have to brake hard, according to what we know from business economics?

In days gone by the situation would have been crystal-clear; it would have been 'full speed ahead'. Provided there was enough money to pay off the interest, and as long as the bank was willing to lend us money, expansion was the rule.

Now the situation is different.

The new board has clearly spelt out that they want to see if OPERATOR can really make money. They are also critical of the company's high interest payments.

According to the suggested actions for sector 5 of the business opportunity matrix, 'overcapacity and a red light', we should avoid the temptation of trying to expand out of a situation where both profitability and financial strength are below target. It is true that we have overcapacity, but to fill it by taking market share we would have to make large investments in system and market development — investments that would be written off in the first year.

If we could somehow find a situation where customers started buying from us without such investments, that would be a different matter. But expansion does require increased marketing assets, which would lead to increased debt.

In this situation the rules tell us to retreat, which normally means retreating and giving up market share.

We can do this by reducing our marketing activities and then, naturally, using the competitive weapon that means least and costs most, i.e. price. We increase prices and then try to adjust our costs to the volume needed, which most likely will entail the hard decision to reduce staff.

But Charles Armstrong was squeezed between a demanding board and an unsympathetic organisation.

How could Charles get 'Brains' and the unions to understand the necessity to reduce costs and dismiss people, having only recently presented a profit of £1.0 million (before provisions and tax; see the annual report in Appendix 1) just to achieve some obscure percentage requirements that had never been mooted before?

STRATEGY
DEVELOPMENT

Economy ◄──

History

Present
situation

Business units

Competitive
analysis

Evaluation of
business units

Prioritising the
business units

Goal setting

Definition of
core questions

Execution

If the annual report had been disastrous, it would have been much easier to get the organisation to accept this bitter medicine. Charles almost regretted both the customary year-end rush to get products invoiced and his high valuation of stock and work-in-progress. But he was unsure of his own position with the new board and dared not reduce the profit further; such options are normally only open to newly appointed managing directors.

For once, the action-oriented Charles Armstrong was counting on help from both above and below — from above in the form of some recovery in the economy, and from Brains' cellar through the next generation of flame control systems.

It was time to harvest previous investments in system development. Charles sat down with the consultant and his Finance Director Malcolm to make some rough calculations.

OPERATOR is, as we have seen, very sensitive to changes in the economic climate. The systems it sells are investments that customers can easily postpone if they are worried about their own liquidity in a recession.

A difference in performance (e.g. sales) between the top and the bottom of an economic cycle of 20% is quite common. The recovery in the trade that should have arrived in 1993 has now begun to be felt, and therefore Charles felt he could count on some sales increase without a corresponding need for increased marketing effort.

| | 1994 | 1995 | 1996 | 1997 |
|---|---|---|---|---|
| | | £ millions | | |
| Cyclical sales change | | +3.0 | 0 | −3.0 |
| Incremental contribution | | 60% | | 60% |
| Change in result | | +1.8 | 0 | −1.8 |
| TOTAL RETURN | 1.9 | 3.7 | 3.7 | 1.9 |
| Change in marketing assets | | +1.5 | 0 | −1.5 |
| Change in free loans | | −0.2 | +0 | 0.2 |
| Change in working capital | | +1.3 | 0 | −1.3 |
| TOTAL WORKING CAPITAL | 18.2 | 19.5 | 19.5 | 18.2 |
| PROFITABILITY | 10% | 19% | 19% | 10% |

Charles Armstrong is a careful and prudent man. He only assumed growth based on the recovery in the economy, something he had seen many times before. He could have calculated with additional sales increases in the next few years from the larger market share that should result from the launch of the new generation of flame controllers, but he refrained from doing so.

'You can be certain that our competitors aren't sitting idle while we develop new systems!' was his comment to a junior director.

He may also have been justified to assume market growth of a few per cent per year, a trend that had been seen for the last few years. All hinges on the recovery in the economy, but Charles is certain that it is on its way. He has been in the business long enough to know that there is a light at the end of the tunnel (and that it isn't an oncoming train!).

He also knows with OPERATOR's cost structure, the results will improve almost automatically provided market share is maintained. This, of course, does require retaining sufficient capacity to be able to deliver quickly.

Thanks to the actions in Department S, OPERATOR is now well prepared for the coming recovery, but the situation has changed in so far as the profits can no longer be spent only on system and market development. The board wants to see improved profitability first. Here, Charles's managerial skills will be sorely tested.

It is not easy to rein in Brains and the other development engineers when there is money available in the company. Charles needs to apply firm control to hold down the capacity costs.

But what worries him most is the financial strength.

The goal is to reach an equity ratio of 40% before attacking. This is how Charles calculated it:

| | 1994 | 1995 | 1996 |
|---|---|---|---|
| | | £ millions | |
| Return | | 3.7 | 3.7 |
| Interest | | −0.9 | −0.6 |
| Management fees | | 0 | 0 |
| Tax | | −1.3 | −1.3 |
| RESULT BEFORE PROVISIONS | | +1.5 | +1.8 |
| EQUITY + DEFERRED TAXATION | 5.0 | 6.5 | 8.3 |
| Increase in free loans | | 0.2 | 0 |
| Amortising | | −2.6 | −3.9 |
| LOANS | 15.5 | 13.1 | 9.2 |
| TOTAL ASSETS | 20.5 | 19.6 | 17.5 |
| FINANCIAL STRENGTH | 24% | 33% | 47% |

STRATEGY
DEVELOPMENT

Economy ⬅

History

Present
situation

Business units

Competitive
analysis

Evaluation of
business units

Prioritising the
business units

Goal setting

Definition of
core questions

Execution

The calculation assumed that investments in fixed assets would be balanced by depreciation.

Charles Armstrong knew that the equity ratio would not develop as well as the calculations showed. He believed in the economic recovery, and therefore in the improved results, but neither he nor the other members of staff would like to hold back on investments in system and market development to the extent the calculations showed.

The danger Charles saw was that Head Office would reduce OPERATOR's profits by charging management fees once the recovery had arrived. The other companies in INVESTOR plc did not have profitability development to justify letting OPERATOR keep its profits — they needed the money more. Charles couldn't find any real arguments for improving OPERATOR's financial strength, if this meant the other companies would have to pay extra tax.

'I can handle the tax man,' he said. 'But it's those bailiffs from Head Office that are unstoppable.'

The consultant had some ideas on how to convince Head Office of the merits of some kind of fund system, earmarking management fees for the future refinancing of OPERATOR, but Charles Armstrong didn't have much faith in them.

'If the money's gone, it's gone,' he repeated stubbornly.

Charles was sorely tempted to 'invest away' money in development during the first year, but his sense of responsibility won and he decided to rein in his warhorses for the first year and try to expand in the two subsequent years. This was all on the condition that the sales development was better than calculated; i.e. that OPERATOR was gaining market share. He would never again repeat the mistake of sizing his capacity after the demand at the peak of the economic cycle.

Charles didn't believe for one moment that things would go exactly as planned, but he had produced a well thought-through argument to avoid panic and the need to lay people off.

We have now established the economic platform for our strategy development. Once we know that no short-term measures are needed to save us from bankruptcy or other calamity, we can continue our strategy development in a methodical fashion.

# What can we learn from our history?

'You're allowed to make mistakes,' one of my first bosses told me, 'so long as you don't repeat them.'

We do not have the time or the energy to make an in-depth description of our company's history. We should, however, recap sufficiently thoroughly to avoid repeating old mistakes and to learn from what we have done well.

## How did it all start?

OPERATOR Ltd. was founded by Sir Peter Byrne and Blake Fieldman in 1975.

It all started because Sir Peter wanted to improve the efficiency of his oil-fired central heating system after the steep increase in fuel prices in the early seventies. On a trip to the USA he had bought a small micro-computer from Radio Shack which he tried to use to control the combustion process.

Sir Peter's knowledge of electronics was insufficient to cope with such a complex system, so he contacted Blake, who worked in the development department of the same company as Sir Peter.

Blake has always been a genius when it comes to electronics and he soon overcame the problems, more than doubling the efficiency of Sir Peter's invention.

## Why?

Rumours of Sir Peter's and Blake's success with the central heating system soon spread, and orders started to come in. Nights and weekends were used to fulfil the orders.

After a while Blake and Sir Peter took the decision to start their own company and work in it full-time. Blake did not have much money and could only put up 10% of the share capital.

Soon they found that the company could not grow by simply catering to the private house market. The really big money could only come from industry.

Sir Peter knew a senior manager in a local foundry: that's where the first industrial application was developed. Their contact had a son who had just finished college and needed a job — his name was Charles Armstrong and he went on to become Managing Director of OPERATOR.

## How has it developed?

For the first few years turnover doubled annually, since when growth has averaged 25% per year (including inflation). It is difficult to estimate the price development, but we can guess at a growth in real terms of at least 20% per year which is probably about the same as the market's own growth.

The conclusion has to be that OPERATOR has been unable to increase its market share significantly in recent years.

| STRATEGY DEVELOPMENT | |
| --- | --- |
| Economy | ✓ |
| History | ◄ |
| Present situation | |
| Business units | |
| Competitive analysis | |
| Evaluation of business units | |
| Prioritising the business units | |
| Goal setting | |
| Definition of core questions | |
| Execution | |

One reason is said to be the lack of financial resources to invest in development — OPERATOR is in a very development-intensive industry. Whether this is true is debatable. It should have been possible to use the market's inherent growth to make the expansion self-financing.

The return on capital employed (ROCE) has been around 12% which has just about paid the interest on money borrowed; i.e. OPERATOR has been showing break-even results and a financial strength of 15-20%.

## Mistakes and strokes of genius

The first couple of years in the central heating market were wasted; there was no money in it. In addition, neither Sir Peter nor Blake had a clue as to how to market directly to consumers.

The attack on the foundry market, however, was a stroke of genius as it enabled development to accelerate dramatically.

The Singapore office should never have been opened, since the company didn't really need it, couldn't afford it and, besides, the competitors were already there. On the other hand, it was fortunate that a subsidiary in the USA was set up before the Japanese entered that market.

OPERATOR has encountered criticism that it has spread itself too thinly in too many applications areas. Even if it is accepted that it possesses unique skills in micro-processor control, doubts exist of its suitability, for example, as adviser in metallurgy.

## The original market segmentation

The first aim was to save energy for British home owners. Thereafter came the start of business in British foundries. When customers wanted systems installed in their foreign subsidiaries, there was a natural drift into exports to other European countries and the United States.

Once the energy crisis was over, profitability became squeezed and additional benefits to customers were required to justify high prices. These focused on improved quality, a side-effect of the improved control of the combustion process.

The ability to discuss quality in foundry applications was something OPERATOR had learnt through experience in the industry over the years. But when it diversified gradually into new applications, things became worse; metallurgists had to be hired, and then cellulose chemists too.

The more complicated the technology became, the more unpaid time had to be spent on preparing projects and so on.

Finally, the decision was taken to split the organisation's markets into hardware and software. This made it possible to charge separately for consultancy and programming.

This original segmentation is described in the diagram below.

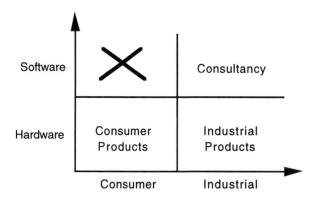

## The corporate culture

Whilst Sir Peter was alive, there was a good atmosphere throughout the company. There was an attitude of fearless attack and, although discussions could get a little heated at times, the company was still perceived as one big family — where everybody was working for the same goal of making OPERATOR big and profitable.

The problems started when Sir Peter died and his heirs wanted to sell out; as Charles and Blake did not have enough money to buy the company, it was sold to an investment company instead. Blake locked himself in the laboratory and didn't show his face for several days. He refused to acknowledge the new board members even though he still had 10% of the shares.

Blake used to be in charge of both production and development but his attitude and lack of co-operation made it necessary to recruit a production manager. He now devotes his time exclusively to development, but has not been able to produce anything truly revolutionary for a long time.

The organisation is currently complaining about all the extra paperwork that the new Finance Director has introduced, after pressure from Head Office.

Charles Armstrong tries to be loyal but when he is reading memoranda from Head Office, his body language reveals that he feels tied hand and foot.

Board meetings have changed drastically. Having once been a discussion between Sir Peter and a chosen few, it has now changed to 'third-degree interrogation' according to Charles. The new board members appointed by the new owners only have the vaguest idea of OPERATOR's business.

This was the consultant's first draft of the history of the company, following discussions with Charles Armstrong and his key personnel. Charles agreed that the description was basically correct but demanded certain adjustments before it was presented.

From this description, Charles realised it must be obvious to everyone that his attitude to INVESTOR was too negative. He decided to set aside his resentment at not having been able to buy the majority share of the company once and for all. After all, INVESTOR is the second-best alternative as owner, after himself. It could have been much worse, since INVESTOR is financially quite strong and has a skilled management team. Charles Armstrong and the consultant were therefore in agreement on several important points.

| STRATEGY DEVELOPMENT | |
| --- | --- |
| Economy | ✓ |
| History | ◄ |
| Present situation | |
| Business units | |
| Competitive analysis | |
| Evaluation of business units | |
| Prioritising the business units | |
| Goal setting | |
| Definition of core questions | |
| Execution | |

1. OPERATOR is in danger of being overtaken by stronger competitors as market share has been slipping in the last couple of years. This is perhaps due to the fact that it has not been possible to make expansion self-financing and partly because there has been diversification into too many different activities.

2. The original driving forces behind the development of the company are almost gone. Sir Peter is dead and so is Blake's motivation. What would happen if Blake resigned or joined a competitor? Charles says that he has taken on much of both Sir Peter's and Blake's roles, but it is clear that he knows this is only partly true.

3. The new board does not want to go on investing with the low profitability OPERATOR has shown so far. One reason for this is the difficulty of evaluating the investments OPERATOR is making in system and market development.

The intention here is not to set out a complete history, but to capture something of the soul of the company and to sort out what is important strategically.

In the case of OPERATOR, the important result is the recognition of the three points mentioned above. In another company it may well be something quite different.

Let us now continue with the next section in OPERATOR's strategy plan.

# The present situation

As we have mentioned previously, analysis of our current situation can constitute a risk of 'paralysis through analysis', as we are diverted from our day-to-day activities that keep the business moving.

There is, however, a tried and tested method to compress the work; using a 'SWOT' analysis (strengths, weaknesses, opportunities, threats). This method concentrates on important changes in the external and internal environments. You may well have come across it already.

It is important to note that it is seldom worthwhile carrying out a SWOT analysis for the whole company. Most likely there are several quite different business units and it is at this level that a SWOT analysis is most useful.

We have carried out SWOT analyses for OPERATOR under two headings: 'What is happening around us?' and 'What is happening within our company?'

## What is happening around us?

The most important event is the fall in the price of energy. This reduces the value of the benefit we can give a customer when saving energy. Additionally, we have halved the remaining market by consultancy services. In many instances, customers have been satisfied with adjusting their production process using manual methods following our analysis.

That the market situation for foundries is difficult is well known but the latest recession has made their financial situation so precarious that even defensive investments like energy control systems have to be postponed.

The one straw that we can clutch at in this sea of trouble is our customers' increased interest in quality. This has resulted in substantial development investments in OPERATOR to increase capabilities and awareness of quality improvements through flame control.

# What is happening within our company?

In our ambition to increase income we are diluting our competence. Apart from the basic competence in energy-saving, which is of reduced value nowadays, we are on the way to forming four 'guilds' in the company. We have four consultancies:

1. The 'old' energy-saving specialists;

2. Foundry specialists;

3. Aluminium specialists;

4. Pulp specialists.

Additionally we have a hardware and component company struggling to fulfil the demands and wishes from the consultancies.

In the absence of a clearly-stated strategic plan to control resources, there is a constant battle between the groups. At times, tempers do boil over.

'I don't know who are the worst,' someone — who wished to be anonymous — said, 'my colleagues or the competitors.'

# Are we still one company?

It is a fair question now to ask whether OPERATOR should be split into two or more companies. It is a purely theoretical question, however, as none of the activities — apart from the original — is big enough to survive on its own.

# A review of our business idea(s)

OPERATOR's original idea was to 'save energy in foundries by using flame controllers'.

It is obvious that we no longer stick to the original business idea. We are now also actively consulting in energy-saving, in metallurgy (within both foundries and aluminium smelting plants) and in the pulp industry.

| STRATEGY DEVELOPMENT | |
| --- | --- |
| Economy | ✓ |
| History | ✓ |
| Present situation | ◄ |
| Business units | |
| Competitive analysis | |
| Evaluation of business units | |
| Prioritising the business units | |
| Goal setting | |
| Definition of core questions | |
| Execution | |

## Our company structure

The OPERATOR company is part of the INVESTOR plc concern. OPERATOR has an independent annual report and its own financial control. The auditors, however, are appointed by INVESTOR, as are all the board members. INVESTOR also reserves the right to charge management fees.

OPERATOR is split into three business divisions called profit centres:

● Consumer products

● Industrial products

● Services

Their profitability is controlled but the accounts do not separate out the capital tied-up in each division. Profitability is not integrated with the overseas subsidiaries; they report directly to the Managing Director and are considered as independent profit centres.

OPERATOR is active in a large number of geographical business areas; in fifteen countries in all. There are wholly owned subsidiaries in Belgium, Germany, Sweden, USA and Singapore.

Not all business divisions are active in all countries, but we can count twenty business units (where a business unit is a combination of a business division and a business area).

This was the consultant's first draft; he realised that adjustment was necessary even before Charles Armstrong saw it. The important thing, however, was the result of the analysis and he made sure this was accepted by both Charles and all his key people.

It is very tempting to start taking strategic decisions in the middle of an analysis of problems. Of course, it is important to capture good ideas but they should simply be noted down without too much discussion until the analysis is completed.

The consultant, for example, was itching to suggest a different company structure. The split between business divisions is no longer rational and it seems obvious that the capital tied up should be shown separately for each business division. Additionally, the overseas sales offices should be integrated with the respective business divisions. Property and financing should be made separate divisions so that a business unit's result can be judged after market-based rent and a yearly fixed alternative interest.

Now that we have considered and established the economic platform for strategy development, i.e. in which square of the business selection we find ourselves, we can continue with the second very fundamental strategic decision, namely how we should regard our market and our business.

| STRATEGY DEVELOPMENT | |
|---|---|
| Economy | ✓ |
| History | ✓ |
| Present situation | ✓ |
| Business units | ◀ |
| Competitive analysis | |
| Evaluation of business units | |
| Prioritising the business units | |
| Goal setting | |
| Definition of core questions | |
| Execution | |

# Business units

The smallest indivisible units in our business are the business deals we have with each individual customer. Were we to discuss ideas at this level, we would never risk losing touch with reality. Unfortunately, however, strategic discussions cannot cover such detail. We must try to find the broader patterns in our business.

Once we have decided on a certain split in business units, we have in effect focused our minds in such a way that it can be very difficult to see business opportunities outside this framework. It is therefore no exaggeration to suggest that deciding how to organise and structure the business into these units is a basic strategic choice which is likely to influence all subsequent decisions.

Take OPERATOR's separation of its business into hardware and consultancy, for example. This philosophy stifled development towards selling systems. It also meant that for several years OPERATOR's unique knowledge and competence were sold at a low hourly rate.

The process of dividing the future activity of the company into business units can be made very theoretical. I believe it is good to be theoretical in the initial stages but to let a certain 'gut feeling' help us to take the final decisions.

We start by using our left brain to go through all the possible business segmentation variables and end up using a creative right brain exercise where we need to derive a maximum of around ten business units.

## A *logical analysis*

Any market can be represented in several ways. The consumer marketeers have the most dimensions from which to choose. They can segment the market according to age, sex, social group, dwelling, hobby, lifestyle, education, height, weight and so on.

In technological business life is somewhat easier. Industrial products and services can generally be represented by just three dimensions: technology, applications and geography, as shown schematically in the following diagram. It should be remembered that other such variables are equally possible — for example, company size, hardware/software etc.

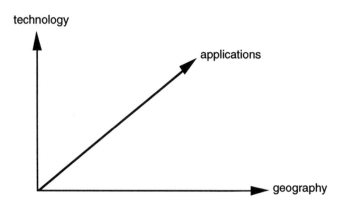

This is how OPERATOR segmented its markets.

## SEGMENTATIONS

### 1. Technology

● Control of combustion processes using micro-computers and sensors.

● Metallurgy — cast iron

● Metallurgy — aluminium

● Pulp chemistry

### 2. Applications

● (Oil-fuelled central heating systems)

● Foundries

● Aluminium smelting plants

● Pulp plants

### 3. Geography

Thirty or more countries are possible as future geographical markets.

More interesting, perhaps, is what is not on the list, namely the additional variable (or dimension) of hardware/software.

Following earlier discussions on pricing policy it had been felt that it was time to integrate hardware and software into system sales again. This way, it would be possible to offer better solutions and, at the same time, widen the base of what could be charged for.

This decision had not been arrived at without a struggle, however. It crossed quite a few boundaries — which was precisely Charles Armstrong's intention.

## A creative synthesis

How many theoretical segments can we find in OPERATOR's case?

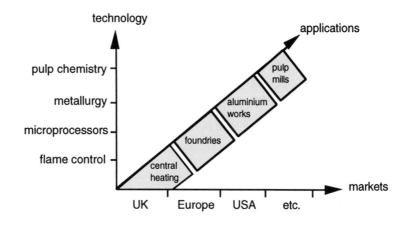

Four technologies x four applications x thirty countries = 480 segments. It is true that some of these segments (boxes) are empty. Few foundries need knowledge of aluminium metallurgy, but it is still quite impossible to work with several hundred genuine business units in our strategy development. We must reduce this down to a maximum of ten.

We can go part of the way using logic.

The number of technologies can be reduced to two. We know already, having assessed the economic platform for our strategy development, that we cannot afford to split metallurgy into cast-iron and aluminium separately. We also know that everyone has to have a knowledge of process control, irrespective of the application; this is, after all, our basic claim to fame.

The split on the technology axis will therefore be simply into metallurgy and pulp chemistry.

The number of applications can, in principle, also be reduced to two. Flame control for central heating systems is a nostalgic remnant we do not intend to develop. This application is only available for our home market. For financial reasons neither will we be able to afford a specific competence for foundries and aluminium smelting plants.

The split on the application axis will therefore be into **Metal** and **Pulp**.

Our economic platform does not allow us to develop more new geographical markets. The number of units along the geographical axis can therefore be reduced to fifteen where we are already established. It is not realistic to open up new markets within this plan.

We have now reduced the number of possible business divisions to sixty (two technologies x two applications x fifteen countries = sixty segments). We are well on our way towards our maximum of ten business units.

## Defining business divisions

The next step on the way is to decide on business divisions — or even division, singular. By 'business division' we mean a global speciality. In Chapter 4 we defined it as 'a strategic unit within a company which can be assigned a reasonably independent balance sheet and profit-and-loss account. It will also control reasonably independent resources to pursue a distinct business idea'.

Ideally I would like to see a company consist of only one business division, but that rarely happens in practice.

When 'divisionalising' the business, the earlier analysis of market segmentation is helpful but the final stages have to be guided by gut feel.

In the case of OPERATOR, it felt right to split the business into two divisions: Metal and Pulp. For familiarity, it was decided to keep the old term 'profit centres'.

Charles Armstrong had accepted the idea that subsidiaries should be seen simply as administrative units and gave Malcolm, the Finance Director, the task of consolidating their results into the two business divisions. The business divisions should also be given a balance sheet. That gave Malcolm more headaches but as the emphasis was on 'be given', he accepted the task. The important thing was the correct allocation of the marketing assets.

Not all the heads of the subsidiaries accepted their new role. Two of them resigned but the rest found their job more meaningful within the new organisation.

| STRATEGY DEVELOPMENT | |
| --- | --- |
| Economy | ✓ |
| History | ✓ |
| Present situation | ✓ |
| Business units | ← |
| Competitive analysis | |
| Evaluation of business units | |
| Prioritising the business units | |
| Goal setting | |
| Definition of core questions | |
| Execution | |

STRATEGY
DEVELOPMENT

| | |
|---|---|
| Economy | ✓ |
| History | ✓ |
| Present situation | ✓ |
| Business units | ◄— |
| Competitive analysis | |
| Evaluation of business units | |
| Prioritising the business units | |
| Goal setting | |
| Definition of core questions | |
| Execution | |

# A business idea for each business division

OPERATOR's business idea could now be stated for each business division. These can also be called 'objectives' in some business analyses.

## Business idea for the METAL profit centre

By using flame control systems, we will, in order of priority:

1. save energy in foundries;

2. save energy in aluminium smelting plants;

3. increase quality in foundries.

## Business idea for the PULP profit centre

With the help of flame control systems we will save energy in pulp mills.

Several hard battles were fought during the selection and definition of these business ideas.

As a business idea is, in a way, a summary of the company's strategy, we can see the outlines of future business prioritisation already.

Choosing strategy in post-industrial business development means discarding a number of alternative activities as the basic strategy is to focus on niche areas.

The descriptions of the business ideas in OPERATOR suggest that emphasis will again be placed on the original activity — saving energy in foundries. It is also clearly stated that it is only in foundries that the company has sufficient competence to address quality problems. In the pulp application area, activity should be concentrated on energy savings.

Behind these decisions lies the realisation of how important it is to concentrate your competence and how expensive it is to develop new applications — not to mention the cost of persuading customers that we have competence in this new application.

## *Defining business areas*

The choice of geographical markets is easy in OPERATOR's case. It is not possible economically to start up in more markets. In other cases, it will be necessary to go through the process described in Chapter 4.

## *Defining business units*

We are now ready, finally, to define our ten business units.

One possibility would be to combine the two business divisions and existing business areas. But it is not that simple!

There are two business divisions in OPERATOR now — Metal and Pulp. As they are both active in fifteen markets, this would give thirty business units.

There are two main avenues we can take to bring this number down to ten or less. One is to lump countries together into larger geographical areas. Another is to lump together the strategically less interesting activities under a catch-all title such as 'Miscellaneous'. This rather anticipates our strategic prioritisation but that cannot be helped. We have to get on with it.

OPERATOR decided to consider the following business units.

### BUSINESS UNITS IN OPERATOR LTD.

```
METAL (UK)

METAL (Scandinavia)

METAL (Europe)

METAL (USA)

METAL (Canada)

PULP (Scandinavia)

PULP (USA)

PULP (Canada)

Miscellaneous
```

You can imagine the discussion that broke out in the organisation when this list was made public!

| STRATEGY DEVELOPMENT | |
|---|---|
| Economy | ✓ |
| History | ✓ |
| Present situation | ✓ |
| Business units | ← |
| Competitive analysis | |
| Evaluation of business units | |
| Prioritising the business units | |
| Goal setting | |
| Definition of core questions | |
| Execution | |

Although a re-definition of business units does not have to lead to changes in the basic structure of an organisation, this time the changes were so radical that they could not be avoided.

The consultant was quick to point out that it was time to forget prestige in the old pyramids and that individuals should be rewarded for the effort they put into projects to realise the new plan.

For once, Leonard Hewson was happy and positive to the changes. He, along with everybody else, realised there could hardly be anyone to challenge him for overall responsibility for the largest business division — Metal.

When it came to Pulp, the situation was more difficult. Would Carol, or her husband, have overall responsibility?

We will, however, save the organisational problems until later. At this point it is time to consider the competition.

# Competition analysis

A strategy that does not take the enemy (i.e. our rivals) into account is of little worth. But all too often we do bury our heads in the sand and pretend that we have no competitors.

'Well, they don't compete directly with us', is something I often hear. 'We are much more specialised.'

How often do you discuss what competitors are doing in your company? Daily? Every month? Once a year?

Although some people will suggest we do not have any competitors worth mentioning, rarely do we sell as much as we would like to. Why is this?

Perhaps it is because it is not our direct rivals who are stopping us from selling more but so-called 'substitution competition' or 'budget competition' (i.e. money being spent on other things instead).

But let us start with what everyone thinks of when they hear the word 'competition' — namely, rivals in our chosen industry.

## Who are our rivals?

Systematic collection of available information about rivals is a great asset, including annual reports, brochures, press cuttings etc.

To get an overview, a file should be collated which starts with the following details:

- Name
- Location of headquarters
- Turnover
- Number of employees
- Parent company
- Profit development
- 'Interesting items'

As an example of an 'interesting item', I would note that Stuart Mills, one of OPERATOR's competitors, is a family-run business but is currently in great financial difficulties.

## What do they do?

The following table gives an excellent overview of the activities of OPERATOR's rivals.

| | BUSINESS UNIT | | | | | Turnover index | | | | |
|---|---|---|---|---|---|---|---|---|---|---|
| | Metal (UK) | Metal (Scan) | Metal (Eur) | Metal (US) | Metal (Can) | Pulp (Scan) | Pulp (US) | Pulp (Can) | Misc | Total |
| OPERATOR | 25 | 10 | 15 | 32 | 3 | 3 | 1 | 1 | 10 | 100 |
| **Rivals:** | | | | | | | | | | |
| Stuart Mills | 1 | 20 | 5 | 65 | 20 | | | | 489 | 600 |
| Amer. Control | | 5 | | 75 | 25 | | | | 95 | 200 |
| Herb. Geisman | 10 | 25 | 75 | 10 | 5 | | | | 175 | 300 |
| Oy Hogman | | | | | | 25 | 50 | 10 | 315 | 400 |
| Kosumura | | | | 20 | 10 | | | | 770 | 800 |
| Our ranking | 1 | 3 | 2 | 3 | 5 | 2 | 2 | 2 | – | – |

The table is reasonably straightforward provided you do not try to be too exact. Concentrating too much on detail will do little to increase the quality of the strategic decisions.

STRATEGY DEVELOPMENT
Economy ✓
History ✓
Present situation ✓
Business units ✓
Competitive analysis ◄
Evaluation of business units
Prioritising the business units
Goal setting
Definition of core questions
Execution

To complete the table, first fill in the 'total' column on the right, starting with your own company on the first line, giving it a total index of 100. Then split that 100 up into the relative proportions, by turnover, of your different business units.

Move on to your biggest rival, giving them an index in proportion to how many times larger or smaller than yours their turnover is. If turnover is not a relevant measurement in your industry, you can use something that describes the scope of the activities better; for example, number of employees.

It is of course more difficult to divide up your rivals' sales into their respective business units but, again, make your best estimates. If it should turn out that a strategic decision will be dependent upon a more exact estimate, we can spend more time and effort on a tailored market survey later.

The 'miscellaneous' column is very important. It should be complemented with notes to explain what it involves that we feel is outside our activities. Often you underestimate the strength that competitors obtain from their activities taken together, although it is also possible to overestimate this.

OPERATOR could read much interesting information from the table and even came up with an idea for a strategic move. The table confirmed that they were too thinly spread and that the strongest competitor, American Control, was doing the opposite. OPERATOR had every reason to fear the day American Control might decide to launch an attack on the European market.

OPERATOR saw a golden opportunity to strengthen its position in North America by buying, or merging with, Stuart Mills. The company was in financial difficulties and there was reason to believe the family would like to concentrate on the foundry business in which the company's origins lay.

## Other competition

Substitutions for what we offer are almost always available. Trying to see the world through the customer's eyes, it is normally possible to make a shortlist of possible substitutions and find ways of meeting the competition from these.

OPERATOR has lost many sales opportunities with customers who have gratefully accepted its analysis and then invested in manual control systems.

Another reason customers turn down offers from OPERATOR is that money is needed for other investments that are considered to be more urgent. There will always be this 'budget' competition. Most customers will already have a limited budget for investment and this is cut even further during a recession. Our suggestions therefore compete with all other types of investment.

If you are not convinced that you can present an investment calculation which is more favourable than all the other alternatives (and even this may not help in a recession), there is every reason to re-think the approach to your business, as we discussed in Chapter 9.

Perhaps it would be useful in this case to consider changing to payment through usage fees or other alternative pricing mechanisms. In this case, the suggestion can be presented as a saving opportunity rather than an investment requirement.

At a strategic level in decision-making, consideration has to be given to potential competition. In the telecommunications industry, the early 1980s saw gradual disappearance of the boundaries between communications and computers; suddenly the likes of IBM were potential competitors.

Having achieved a good overview of the competition, we can now proceed to evaluate our business units. We are faced with a third main question in our strategy development: where to attack to win market share?

Do you remember the other two?

The first was about our economic platform for strategy development — in which box of the business selection matrix are we to be found?

The second concerned our view of the business; this decides which business units we will discuss in the continued strategy process.

To summarise, our main questions are:

1. Can we borrow more money if we need to?

2. How shall we view our business?

3. Which parts should we develop?

| STRATEGY DEVELOPMENT | |
| --- | --- |
| Economy | ✓ |
| History | ✓ |
| Present situation | ✓ |
| Business units | ✓ |
| Competitive analysis | ◄ |
| Evaluation of business units | |
| Prioritising the business units | |
| Goal setting | |
| Definition of core questions | |
| Execution | |

We have already answered the first two questions so we now proceed to analyse the third. It has to be answered in several stages; first we have to evaluate our business units with respect to market outlook and business position, thereafter we rank them and finally we set objectives and allocate resources.

# Evaluation of business units

Evaluation of our business units is the very basis for sound prioritisation. So as to be convinced of the logic of our evaluation, we can consider it in three separate steps.

First we will evaluate the business units through the tried and tested SWOT analysis (strengths, weaknesses, opportunities, threats). Thereafter we will scrutinise the business idea for each unit and, finally, we can then rank the business units based on their market outlook and business position.

## Strengths, weaknesses, opportunities, threats

To describe our environment completely would be very time-consuming, so we will concentrate on the major changes expected or possible in our business surroundings and class them as threats or opportunities, along with the probability that each will occur.

Even with this structure, the analysis will take time. Eight forms like the one below have to be completed, one for each business unit.

| | Business Unit — METAL (USA) | | |
|---|---|---|---|
| | OPPORTUNITY/THREAT | | |
| Event | Probability | Opportunity | Threat |
| $ up by 10% + | high | x | |
| $ down   5% + | low | | x |
| Energy price rise | medium | x | |
| Energy price fall | low | | x |
| Switch to electric heating | high | | x |

Next we consider the company's internal conditions; these are described by a form specifying our strengths and weaknesses:

STRATEGY
DEVELOPMENT

| | |
|---|---|
| Economy | ✓ |
| History | ✓ |
| Present situation | ✓ |
| Business units | ✓ |
| Competitive analysis | ✓ |
| Evaluation of business units | ◄ |
| Prioritising the business units | |
| Goal setting | |
| Definition of core questions | |
| Execution | |

| Capability | Business Unit — METAL (USA) STRENGTHS/WEAKNESSES | | |
|---|---|---|---|
| | Importance | Strength | Weakness |
| Ceramic cathodes | high | x | |
| Process knowledge | high | x | |
| SPC/RS232 | medium | x | |
| Service (total) | medium | | x |
| Service (New England) | low | | x |

## *Evaluation of the business idea*

The task of evaluating the business idea for each individual business unit is eased as it is normally enough to identify the idea at a divisional level. Differences between the business areas should be insignificant in comparison.

'A good business idea should smell of money,' someone once said. It tends to do so if:

1. The **need** it is supposed to answer is growing;
2. Our **solution** is new;
3. Our **competence** is unique.

Given this definition, what do you think of the business idea for the Pulp business division in OPERATOR?

```
'Using flame controllers, we will save energy in
pulp mills.'
```

Actually it may not be easy for you to say anything, as you are not in the industry.

The **need** to save energy is great, but is it growing?

The **solution** — to save energy using microelectronics and sensors — is no longer new.

Possibly we can claim to have a unique **competence** in this particular solution but our competence in pulp production is weak with only one employee really trained in the industry.

STRATEGY
DEVELOPMENT

Economy          ✓
History          ✓
Present
situation        ✓
Business units   ✓
Competitive
analysis         ✓
Evaluation of
business units   ◀
Prioritising the
business units
Goal setting
Definition of
core questions
Execution

It is not strictly necessary to score top marks in all three areas to make the business idea worthwhile, but the descriptions above did not make Charles Armstrong reach for his wallet.

## Business position and market outlook

What we have done so far can be seen as a dress rehearsal for the final evaluation of our business position and the market outlook for the different business units. These will form the basis for the ranking of which business units to prioritise in the next few years.

Our **business position** should be based on the following criteria:

- **Relative market share**
- **Relative growth** of the market share
- Patents
- Technological advantage
- Capacity
- Image/reputation

The reason I have stressed relative market share and relative growth of the market share is that these should be the dominant factors for judging our position. For the other factors to have any value to us they need to be allied to a high relative market share or knowledge that will grow faster than competitors' so that, sooner or later, we can become 'biggest, best and most beautiful' in the market.

The **market outlook** should be judged using the following criteria:

- **Relative market growth**
- Market size
- Closeness to the market
- Fluctuations in the economic climate
- Seasonal fluctuations
- Price sensitivity
- Expected technological developments
- Barriers

Relative market growth is critical to market outlook. It is defined as '**market growth** divided by **overall capacity build-up**'.

Compare the prospect of being successful in a market with strong growth and where competitors have sold out with no plans to increase capacity, with the situation of a market which has been stable for many years and where all suppliers have considerable overcapacity.

A situation where capacity is growing faster than the market may not be attractive but is still preferable to a market that has already stagnated. After all it is easier to attract a new customer not yet 'in bed with a supplier' than to make a customer leave a supplier where relationships are satisfactory.

Close geographical proximity to the market (or the reverse) also has a certain impact. It is problematic for Pulp (Canada) that the competence is based in Britain.

For the post-industrial company with high fixed capacity costs, all types of demand fluctuations are difficult. Seasonal fluctuations may be acceptable but a drawn-out recession can be fatal.

The post-industrial company tries to avoid price-sensitive markets; its fundamental strategy requires scope for price differentiation.

Expectations of high levels of technological development can be both positive and negative, depending on our situation. If we are a small company with few resources and would need to make very high development investments, it may be questionable as to whether we should enter a market.

Market barriers can, equally, be either positive or negative, depending on our own strength. The knowledge that a multitude of small companies can easily come in and copy us is, of course, discouraging but perhaps we are just such a company. Trade barriers are still a reality, although many of the traditional problems are slowly being overcome.

The list of criteria for judging the market outlook can be made much longer but the first, relative market growth, is so dominant that it is not worth spending excessive amounts of time on the others.

Another reason to stick to relative market share and relative market growth as the main selection criteria is that they are difficult to manipulate; either we are larger and grow more quickly than our competitors, or we don't.

| STRATEGY DEVELOPMENT | |
|---|---|
| Economy | ✓ |
| History | ✓ |
| Present situation | ✓ |
| Business units | ✓ |
| Competitive analysis | ✓ |
| Evaluation of business units | ← |
| Prioritising the business units | |
| Goal setting | |
| Definition of core questions | |
| Execution | |

Even if it is just as big a challenge to bring in money when 'harvesting' a business unit, most ambitious employees seem rather to prefer working in business units which have been given resources to increase market share.

In OPERATOR, Carol's husband argued strongly for Pulp (Canada) by stressing the importance of market size and the good image of British industry.

We will return to the importance of making heroes and heroines out of those whose task it is to create the ambience for improvement. These are the people who are responsible for bringing in enough money to afford the investments needed in market and system development.

If we have managed to complete the SWOT forms, the task of ranking the business units will have become fairly straightforward.

# Prioritising the business units

## *A first ranking*

The evaluations we derived, based on the completed SWOT forms and the analysis in the previous section, can now be summarised as in the following list below. It should be straightforward to rank the business units for each parameter.

| Business unit | RANKING | |
|---|---|---|
| | Market outlook | Business position |
| METAL (UK | 8 | 1 |
| METAL (USA) | 1 | 2 |
| METAL (Scandinavia) | 1 | 4 |
| METAL (Europe) | 4 | 3 |
| METAL (Canada) | 6 | 5 |
| PULP (Scandinavia) | 5 | 8 |
| PULP (USA) | 3 | 6 |
| PULP (Canada) | 2 | 7 |

In order to combine the parameters, we can also now plot the business units on a schematic diagram.

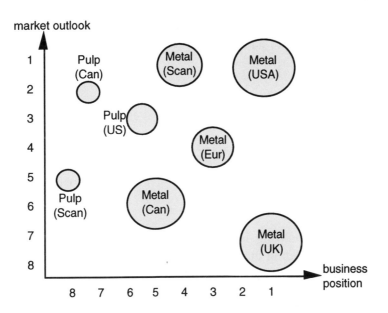

To make our selection easier, we have drawn the circles so that their areas are roughly proportional to the present turnover of the business units.

The first business prioritisation or ranking is almost automatic. The rule is to invest in units in the top right-hand corner first and the bottom left-hand corner last.

The logic is simple. In trying to become 'biggest, best and most beautiful' in a market niche, it is natural first to attack in a niche where you are already 'big, good and beautiful'. The only exception I can think of should be where we already are the 'biggest, best and most beautiful'.

We can also agree that starting up any new operation requires that the market outlook is good.

With this logic, OPERATOR produced the following first ranking:

# Mechanical prioritisation

1. METAL   (USA)

2. METAL   (Europe)

3. METAL   (Scandinavia)

4. METAL   (UK)

5. METAL   (Canada)

6. PULP    (USA)

7. PULP    (Canada)

8. PULP    (Scandinavia)

## *Effects of synergy*

Charles Armstrong is suspicious of what he calls 'mechanical' decisions taken in business development; that's why he insists on labelling this first ranking 'mechanical'.

He is of the opinion that sophisticated points systems, with weighting of one type or another, lend a false sense of scientific exactness when in reality it is mostly a question of subjectivity and guesswork. I also believe that he does not want to be caught in a system that denies him the opportunity to have the last word.

I share his fear of leaving out a decisive factor in the evaluation. One such factor could be the effect of synergy between different business units.

There may be other reasons to deviate from the first ranking. For example, government restrictions in some countries may prevent us from growing too much there by invoking various monopoly (i.e. anti-competition) rules.

## *A revised ranking*

OPERATOR found that the split between 'USA' and 'Canada' was impractical given that the same customers could very well have plants on both sides of the border. Their competitors were also the same. These two business areas were therefore amalgamated to become 'North America'.

Charles also found it wrong to neglect the home market. Even if our market share is very high and hard to increase, it is too early to harvest in the UK. It wouldn't look good in other markets if we were losing ground at home!

The Metal (Europe) business unit is really a miscellaneous group of activities. Diversification within the business unit has been too great to develop it as a unit.

These additional points of reasoning resulted in the following revised ranking:

| STRATEGY DEVELOPMENT | |
| --- | --- |
| Economy | ✓ |
| History | ✓ |
| Present situation | ✓ |
| Business units | ✓ |
| Competitive analysis | ✓ |
| Evaluation of business units | ✓ |
| Prioritising the business units | ✓ |
| Goal setting | ◄ |
| Definition of core questions | |
| Execution | |

## RANKING OF BUSINESS UNITS

1. METAL  (North America)

2. METAL  (UK)

3. METAL  (Scandinavia)

4. METAL  (Europe)

5. PULP   (North America)

6. PULP   (Scandinavia)

There was still no unanimous agreement on this ranking. On the contrary, discussions were quite heated at times. But after Charles had listened carefully to his colleagues' views, he made the final decision which resulted in the ranking shown here.

# Goal setting

We now know what to invest in, but not how much. The goals have to be balanced against the resources available.

## *Market share, capital employed and profitability*

Increasing market share means investment in two areas. First, we have to do more system and market development and, second, we have to inject more capital into marketing assets.

If we cannot make our expansion self-financing, which few will manage to do, we have to create a balance between the business units in which we invest and those where we harvest.

In order to achieve this, Charles Armstrong issues budget directives. This way he ensures that the budget process is linked to the strategic development process.

## Budget directives

Being the careful manager that he is, Charles Armstrong prefaced his budget directives with a moratorium on recruitment and borrowing. The last thing he wanted to consider was having to ask the board for permission to borrow more money. The odd, occasional exception or two could be made to the 'no-recruitment' rule, but he did not intend to let the organisation know this early on in the process.

To say that the company had to be restrained did not mean that everyone had to reduce their activities. Strategy planning is aimed at re-distributing resources in favour of the highly prioritised business units.

# Definition of core questions

We now know where we should invest and where to harvest. But the plan has to be executed too.

A whole string of questions has to be answered before we can think of reaching the predicted market shares whilst, at the same time, respecting the restrictions given. Some of these questions are vital and it is wise to list them before implementing the strategy.

## Listing the core questions

OPERATOR listed the following 'core' questions as being of key strategic importance:

- How can we ensure satisfactory capacity in Department S?

- Do we have 'key-man' insurance for Blake Fieldman?

- How do we approach Stuart Mills?

- How do we obtain acceptance of our strategic plan by Head Office?

| STRATEGY DEVELOPMENT | |
|---|---|
| Economy | ✓ |
| History | ✓ |
| Present situation | ✓ |
| Business units | ✓ |
| Competitive analysis | ✓ |
| Evaluation of business units | ✓ |
| Prioritising the business units | ✓ |
| Goal setting | ✓ |
| Definition of core questions | ◄ |
| Execution | |

The economic basis for our strategy is to be able to use our existing overcapacity in the growing market that will accompany the economic upturn we expect as the recession ends. If it is shown that the bottleneck in Department S will remain, the chances of our strategy being successful will be severely restricted.

The discussion about 'key-man insurance' is a code word for a larger problem — namely, how to get Brains' development department to work. Blake Fieldman (Brains) has become more difficult over the years. He is almost paranoid and will not delegate responsibility to his staff, which results in general paralysis when he is absent. To discuss key-man insurance is a diplomatic and flattering way of approaching the problem, but it is obviously not enough.

The real strategic coup would be to co-operate with Stuart Mills or, preferably, to buy them.

It is wise to make the board (and possibly those who run the board) aware of the strategic plan as the annual accounts are an uncertain measurement of success in post-industrial business development. In the case of OPERATOR, the ability to finance its long-term business development will largely be dependent upon how much Head Office wants to charge in management fees.

In reality, the list of key questions can be much longer than this, but since the aim here is only to demonstrate the method, we will leave the rest to Charles Armstrong and his colleagues.

## Decisions

Decisions to answer these individual questions can only be taken during operational planning and execution. The decision to start,

STRATEGY
DEVELOPMENT

| Economy | ✓ |
| History | ✓ |
| Present situation | ✓ |
| Business units | ✓ |
| Competitive analysis | ✓ |
| Evaluation of business units | ✓ |
| Prioritising the business units | ✓ |
| Goal setting | ✓ |
| Definition of core questions | ✓ |
| Execution | ◄— |

however, can be taken during the strategy development phase.

At OPERATOR it is decided that the manager of Department S should draw up a contingency plan in case capacity problems arise again.

The manager of the American subsidiary is asked to have lunch with one of the members of the family that owns Stuart Mills.

The implementation of the strategy is beginning in earnest.

# Making it happen

Unless you carry your plans through, they are not worth the paper they're written on. Or, to quote a managing director explaining his company's success:

'Our strategies are no better than anybody else's — the difference may be that we carry them through.'

Realising that execution is the most important action, it may seem strange to conclude this chapter with a very short section on this phase. But, on the other hand, execution tends to be something you get on with and do — not write or theorise about!

## *Communicating the strategy*

During the strategy development itself, not much will have been written, mostly just notes and checklists.

But more people will be involved in execution of the chosen strategy than just those who have participated in the strategy exercise itself. The chosen strategy has to be communicated and it is crucial to do it efficiently. In certain critical situations it may even be advisable to involve your advertising agency in communicating the strategy; we cannot afford unnecessary misunderstandings within the organisation.

## Feedback

As communication should be two-way in a modern organisation, there has to be scope for feedback and adjustment.

## The organisation

Very often strategy development leads to changes in the organisation but this is not an essential result.

Changing the basic organisational structure periodically always creates feelings of insecurity and sometimes even paralysis.

As most organisational diagrams rarely live to see their second birthday, a viable alternative to restructuring the organisation could be some form of project organisation. Why not leave the basic organisation as it is and carry out the new strategy in project form?

## Resistance to change

There are always people in organisations who resist change. Resistance to change can have many causes; insecurity, fear of 'demotion' etc. being amongst them.

The best way around this problem is to involve the employees in the strategy development. If that is not possible, it is necessary to communicate, communicate and communicate. We have to sell our new strategy to our staff.

Those who still resist change after all this may simply have to be moved out of the way!

| STRATEGY DEVELOPMENT | |
| --- | --- |
| Economy | ✓ |
| History | ✓ |
| Present situation | ✓ |
| Business units | ✓ |
| Competitive analysis | ✓ |
| Evaluation of business units | ✓ |
| Prioritising the business units | ✓ |
| Goal setting | ✓ |
| Definition of core questions | ✓ |
| Execution | ← |

STRATEGY
DEVELOPMENT

| | |
|---|---|
| Economy | ✓ |
| History | ✓ |
| Present situation | ✓ |
| Business units | ✓ |
| Competitive analysis | ✓ |
| Evaluation of business units | ✓ |
| Prioritising the business units | ✓ |
| Goal setting | ✓ |
| Definition of core questions | ✓ |
| Execution | ◄ |

## Budgeting

Once all the key people have had their say, the budget directives will be drawn up and the budget work starts off in earnest.

With a well thought-through and properly communicated strategy, the work on the budget will be much quicker than would otherwise be the case.

## 'What-if' analysis

The only thing we know for sure about our plans and budgets is that they will not come out exactly as predicted. Before we finally decide on our plans, we have to make sure we can land on our feet if things turn out to be different from our expectations.

It is impractical to calculate 'worst case/best case' scenarios manually but with modern computers and spreadsheet programs, there is really no excuse for not doing these.

Once OPERATOR's goals are fixed for each business unit and the resources distributed accordingly, Malcolm Stone, the Finance Director, will perform a 'what-if analysis'. He will feed the figures into his computer model and predict the profit-and-loss statement and the balance sheet, not only if everything comes out the way they hope, but also in the worst-case scenario. If this worst case scenario proves to be disastrous, more adjustments will have to be made.

## Control and revision

As we know that reality will not follow our plans, we have to build in follow-up and adjustment opportunities in our plans to be used if necessary.

The strategy group should not be disbanded just because the strategy plans have been finalised. Each quarter the group should see a comparison of the plans with reality and be given a chance to analyse any discrepancies. Once a year the whole strategy should be revised and every third year it should be completely re-worked, unless circumstances have forced us to do this earlier.

# Summary

Strategy development is difficult but essential. Certain decisions have greater scope than others and if we do not take them in the proper fashion someone else will do so for us — our competitors, or possibly just chance.

It is important to avoid 'paralysis through analysis' and quickly to get to the core strategic questions. This is done initially by basing the strategy on hypotheses. When the strategic plan is ready, we can test those hypotheses that are crucial for the success of the plan.

But before plans are drawn up, the economic platform has to be assessed. There is obviously a great difference between a plan aimed at avoiding bankruptcy and one geared towards attacking the market aggressively in order to increase market share.

We cannot afford the time to write the complete history of the company but we should analyse key situations from our past and learn from those experiences.

Post-industrial business development is based on focusing on well-chosen niche markets. The way in which we choose these niche markets is vital to the strategy. Once we have decided on the structure of our business units, it can be difficult to perceive activities outside this framework.

Having selected a maximum of ten business units, it is time to evaluate which ones to harvest and which to develop. As market and system development entails investments that are likely to be written off in the first year, we have to find a balance between harvest units and development units. How this balance looks depends on the economic situation.

Well-planned strategy development will, as a by-product, also facilitate and improve the budget work which is normally the main control instrument for business development.

Chapter 13

# Organisation development

*In this chapter we consider how senior management can carry out
its most important role, namely to fill the company with
Business Engineers and give them the right environment
in which to work.*

To invest in BUSINESS ENGINEERING is a major decision. It requires
a willingness to change and focused effort to develop the organisation.
We are changing the old industrial culture and replacing ways of doing
business that have been developing for decades, if not centuries.

After having worked through the previous twelve chapters together, I
hope that we are in agreement that the benefits of such investment are
potentially very large indeed.

It is not a question of turning the company upside down — the change
can be gradual, taking place in many small steps, ideally each one
paying for itself.

## An efficient industrial culture has earned an honourable retirement

*Business decisions based on calculation*

In a company applying a mass-marketing strategy, with cost-based
pricing and cost-based business selection, a university graduate in

economics can take the two most important decisions for the company. Those two decisions are:-

● What should our prices be?

● Which business deals should we accept?

He or she need never have met a customer, does not need to understand in detail what the company sells and can be blissfully ignorant of what its competitors offer. All that is needed is a computer equipped with a suitable spreadsheet or other program for calculating and allocating costs.

## *A strong and successful company culture*

In spite of my negative tone in the opening paragraph, I must stress that I am impressed with the efficiency of that old strategy. It is more or less self-governing and creates structure in business. The decisions are based on solid calculation, not on suppositions and 'good ideas at the time'.

I advise all my consultancy clients to stick to the old strategy provided they can sell on price, and as long as they have flexible costs.

# Cultural evolution

When we are forced to leave the old strategy, it is time to start a cultural evolution. Note the spelling here; there is no 'r' — we seek evolution, not revolution.

Few companies will benefit from a change in corporate culture overnight. It has to be changed, but only with care. We must leave the old bureaucratic corporate culture and move towards one based on business awareness.

This must not be interpreted as a signal to dismiss or reduce the administrative functions, which are still of great importance for the competitiveness of the company, but they cannot play the same controlling role as before. That role has to be supportive, not one of control.

# Your comments

*What is your company's culture? Is it full of post-industrial business awareness? Has it delegated the balance sheet down to business division level? Do all the key people share the same business-economic frame of reference? Are market-based pricing and market-based business selection practised throughout the organisation?*

*If this is the case, your company is unique.*

# Wanted — evolution leaders!

*The business leader — the motor in the process of change*

Shifting a company's strategy to post-industrial business development through Business Engineering requires somebody senior in the organisation who is devoted to this concept; someone who can see the importance of changing the culture. It requires a 'Business Engineering Leader'.

You might well think that the managing director would be the natural business leader; an MD could hardly have a more important role to play than the development of the company's business. Unfortunately most managing directors are very tied up with the day-to-day running of their company, making it difficult to have the necessary time available even if they want to. Without the support of the MD, however, we can hardly achieve change.

# A company's most important resource — the employees

In the first chapter, we established that one of the major changes in business conditions in the post-industrial society is the new position that employees have in the company.

We see that today's knowledge-intensive companies have begun to express some of their key numbers in balance sheets and profit-and-loss statements that are related to employees.

*Internal marketing — as important as external?*

The marketing concept, so important to the latter day industrial society, will survive as a fundamental ingredient in the information or post-industrial society. The company's business planning has to be based on the requirements of people; this is of ever increasing importance today.

When discussing the concept of 'marketing', it is often thought to be directed exclusively at customers.

This is quite natural given the importance of the customer, but as employees have now achieved such a significant position in the company, it may be relevant to use the marketing concept for the staff as well.

An exhaustive discussion of the subject of 'internal' marketing is beyond the scope of this book. I will just touch upon one most important aspect, namely the exaggerated belief in the need for monetary rewards as incentives for employees.

## Abraham Maslow's pyramid of needs

Psychologists tell us that human behaviour is governed by our needs. One famous psychologist, Abraham Maslow[1], has described human needs in the form of a pyramid.

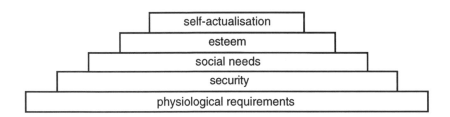

At the base of the pyramid are the physiological needs — drink, food, sex and shelter. If these are not satisfied, the whole pyramid will collapse. If these needs have not been fulfilled, people will jeopardise their security, which is something we have all come across.

Once we have been fed and sheltered, we need to feel secure. Anyone who has visited a company undergoing reorganisation, at a time when the employees do not know whether their own 'box' within the organisation diagram will be maintained, will have seen how low morale can be and how irrelevant social ties become.

In contrast, when an organisation is functioning normally, good contact with colleagues is generally cited very highly by employees.

Sociologists often say that the primary group for the herd animal *homo sapiens* is the family unit, whilst the work group is second. For many

---

[1]Maslow, A. H., *'Motivation and personality'*, Harper & Bros., 1954.

people climbing the career ladder it is usually the other way around. This causes trouble both for the individual and the company.

Having been accepted as a member of a group — perhaps the management team — esteem comes next.

When you are a well-fed and esteemed member of the management team, it is time for the final level in the pyramid — self-actualisation, or self-fulfilment. To develop oneself and feel one's talents grow gives immense pleasure to most people.

## Perhaps money isn't everything after all

'Put them on a commission, then they will really work!' is a suggestion that is often heard, although 'performance-based' and 'productivity-related' remuneration are becoming the more widely used terms. But people do not only work for money. Once the physiological requirements have been fulfilled they want to climb the hierarchy of needs.

## What does a higher salary mean?

In industrial business development (as opposed to the post-industrial concept I have described), the competitive weapon price played a very strong role in external marketing, often to the extreme degree that competitiveness was optimised for price alone. A bare minimum of resources was spent on the other competitive weapons.

When discussing internal marketing — in those instances where it was discussed at all — it will have been natural to assume that the most important means for ensuring the full co-operation of employees would also be through money.

My experience of salary negotiations is that the worst mistake you can make is not to pay someone too little in absolute terms, but to make relative mistakes.

'I get less than she did,' sobbed one of my staff angrily and promptly resigned, without having another job to go to.

What conclusions can we draw from this?

It may be that today's employees are unlikely still to be on the first step of Maslow's pyramid. Perhaps salary is seen more as a sign of the company's appreciation of its employees than as a means purely to increase their standard of living.

## *What can we learn from the army and the Catholic church?*

When I first studied organisation theory, it was said that we had stolen our first organisational structures from the army and the Catholic church. But it seems we have failed to learn how to make people totally devoted to the organisation without a thought for the money it brings them.

What makes people spend large amounts of their time in voluntary defence organisations? What is the attraction of becoming a T.A. officer?

Perhaps one answer can be seen in the very visible way that military personnel are shown appreciation. Put simply, stars are sewn onto their uniforms if they have performed well. (Of course there are many other factors too!)

In the American musical '*Stop the world*' there is a scene which shows the delight of a new board member when he receives the key to the directors' toilet. And in the majority of companies, the size and furnishing of offices are still decided more on grounds of position than need; the art of satisfying human needs by means other than money has not entirely been forgotten.

But much more can be done.

> Back in OPERATOR, if Charles Armstrong had seen the Singapore activity as an internal marketing effort labelled 'interesting and developing jobs', he would probably still have three outstanding engineers on his payroll, complete with all their knowledge of customers and systems. Instead, two of them went to a competitor taking with them business worth more than £1 million in contribution.
>
> 'But surely it can't be right to turn OPERATOR into a playground for globetrotters?' he would complain if anyone broached the subject.

No, of course not — this is an extreme case. However, a modest budget for personal development could actually reduce total personnel costs overall.

Instead of competing just through offering high salaries, could an alternative at the interview be: 'No, it's true that we can't offer the highest salaries in the industry, but we can offer the most interesting and rewarding job in the industry. Additionally, we have a very agreeable workplace and you are welcome to talk to whomever you like amongst your future colleagues'?

The conclusion is that we have to aim at the whole pyramid of needs to attract skilled employees.

## The importance of keeping staff

An important task for internal marketing is to retain the employees we have already recruited. As we saw earlier, it can easily cost £10,000 to find the right new person.

During the first year, the employee is probably only 50% productive so, in addition to that initial investment, we should add another half-year's salary to the cost. Add to this the training — both direct costs and indirect costs in the form of lost time for the 'trained' staff who take on the new recruit — and it is easy to see that the investment could well top £100,000 in the first year.

The longer the employees stay, the more they increase in value to the company and the bigger the loss if we lose them. It could well be worth investing time and money in internal marketing to keep the people we have recruited.

## We need our people to try harder

It's not enough for people simply to stay in the company. We have to make them do their best as well. The difference in contribution between a motivated employee and an indifferent one is huge.

## The secretary who threatened to do exactly as I said

When, in the early days of my career, I tried to set the salary for my secretary, who was getting close to retirement age, the Personnel Manager told me to give her only the absolute minimum increase.

'There's no risk of her leaving and we need the money to keep the younger secretaries with us,' was his explanation.

Young and inexperienced as I was, I took his advice. When my secretary was told of the increase, she turned red with anger and voiced the following threat:

'Just you look out, from now on I'll be doing exactly what you tell me!'

Until then she had always stopped me in time when I was about to do something I shouldn't. She had given me friendly hints of things that should happen but now, if she wanted to, she could just sit and wait for her retirement without performing this vital function. I could hardly force her to use her own initiative and there was nothing that I could do if she started typing at half her normal speed!

The following year, her last but one in the company before retirement, she received a handsome rise.

## From lone ranger to teamwork

There was once a small engineering company helping large pulp mills to cut logs into chips as the first step in the paper-making process. In those days, chipping was done by pushing the log against a large wheel full of large steel knives.

'Tim, the travelling salesman' was selling such knives to the pulp mills. He was always impeccably dressed and had pleasant manners. This was all it took. A deeper understanding of product or application was hardly necessary — it was enough simply to stress the excellent quality of the steel and the exact honing of the knives.

Then, one day the small company had an innovative idea.

The heavy and unwieldy knives were replaced by a cassette system into which small turnable knives were fitted. This ingenious system had a range of advantages and it dramatically improved the quality of the chips, which resulted in better control of the final stages of the paper-making process.

To be able to explain the new advantages fully, it was necessary to have substantial knowledge of cellulose chemistry, which meant Tim was out of a job.

The cassette system had to be especially adapted to each customer's equipment and situation. As time passed, the company developed a sophisticated CAD-CAM system in order to be able quickly to customise solutions for customers.

To sell a system the sales engineer needed to co-operate with a cellulose chemist and a designer. In order to deliver quickly to the USA, the logistics people also had to be intimately involved. If the installation was large, both the Marketing Manager and the Finance Director would have to be involved in the negotiations.

Simple sales had developed into complex teamwork.

## Individual commission is a threat to teamwork

Traditionally, a large portion of the salespeople's remuneration was made up of commission on turnover, without regard to the price level achieved or the profitability.

As we have previously noted, a system of commission works well when Tim the Travelling Salesman sells, say, simple steel knives but less well for system sales.

In Tim's case, almost all the variables in the sales process were held constant. Tim had to sell as many knives as possible at set prices. If he kept sober and made enough sales calls, he sold his quota.

With systems sales, on the other hand, it is by no means certain that the sales engineer is the crucial person; it could equally be a designer finding a particularly good solution for a particular customer, or it could be the quick reactions of the logistics department that makes the difference.

## The sale that never should have been made

When one of the American sales engineers passed a customer's site very late one night on his way to his motel in Vermont, he noticed that it was as light as day around the machine building where the company's knife system was to be installed. The reason for this was that the wood used contained large amounts of silicon, causing sparks to fly continuously when the knives hit these hard particles embedded in the wood.

'Wow!' he exclaimed. 'We shouldn't install our system here — it'll cost us a fortune in guarantee costs!' He knew that the sensitivity of the new system would be problematic under such harsh conditions.

But he still sold the system.

To expect him to turn down the sale would have been unrealistic. His commission wasn't influenced by guarantee costs — and, besides, he had a wife, a young family and a mortgage to pay.

## Profit-oriented commission — a blind alley

As we saw in Chapter 10 for external agents on commission, one of the big drawbacks with the old-fashioned commission system for salespeople is the creation of internal price sensitivity. Commissioned salespeople, although fully employed by the company, are the worst enemies of price increases, just like commissioned external agents.

And how could it be any other way?

Assume we contemplated risking a 10% price increase to double our net margin. How is this risk analysed from the salesperson's point of view?

If we increase prices by 10%, the salesperson has 10% to win, but 100% to lose. If he or she has to try twice as hard to sell the products, a price increase is clearly a bad proposal from his or her point of view.

Some companies have identified this problem and tried to solve it by basing commission on some sort of profit measurement - e.g. contribution instead of turnover. The problem here is that there is no easy measurement of profitability that will give the desired result.

In this example, if the salesman's commission was based on the contribution of the American business, there would be an endless discussion about internal pricing between the European parent company and the American subsidiary.

The basic problem is that it is no longer possible to pinpoint a single individual as responsible for a successful result. Even the most sophisticated and individual commission system does not adequately reflect this modern style of business.

## How should we do it now?

It is easy to criticise something, but what can we put in its place?

I agree with those who say that companies have to increase the incentives for their employees by sharing the fruits of success, but it can be achieved in a better way than through commission on turnover paid out to a few individuals.

As we have found that success, to an ever-increasing degree, is the result of effective teamwork, it would seem logical to reward groups instead of individuals. And who says that this has to be paid out in money? Is it not possible to have a better effect through other means of rewards?

Is it possible to stop having individual incentives altogether?

Probably it is not, but they should be based on individual judgements. Management cannot avoid spending time on the individual to find out how well he or she works. As team leader you cannot leave your staff to fend for themselves out in the field.

If the sales engineer up in Vermont had taken the decision not to sell, he would have been worth a salary increase instead of a decrease. Overall, the loss of one sale to the company is of small consequence compared to the potential costs of fulfilling guarantee promises.

I think that just having the knowledge that management looks at their work closely enough to consider whether it is worthy of a salary increase is enough for today's people to try harder.

To sharpen the argument, I would say that the problem today is not one of employees in key positions being lazy; rather it may be that they try too hard and in so doing can ruin the overall teamwork.

Outstanding individual accomplishments should be rewarded, but when individual accomplishment and the overall results are not closely linked, there needs to be more than just slotting the individual's accomplishments into a commission system.

# A post-industrial business development programme

When I considered the title of this section, I hesitated between 'Organisation development programme' and 'Business development programme'. I soon found that it does not make much difference. Development of a post-industrial business is intimately tied to developing the organisation and its personnel.

## Business development potential

It will cost money to carry out a post-industrial business development programme but that cost should be related to the huge potential benefit from such development work.

Let us take OPERATOR as an example.

The area with the fastest pay-back is pricing according to customer-perceived value (CPV). Leonard Hewson estimated that by spending resources on a more carefully thought-out pricing system, the average price level could be increased by 10% for half the business deals. With a turnover of £33 million, this would result in an increase of approximately £1.5 million.

OPERATOR's present situation is one of overcapacity. A market- and capacity-oriented business selection strategy could increase capacity utilisation by at least 10% which, at an incremental contribution of at least 30%, would result in an increase of £1 million.

OPERATOR has been far too generous with financing, despite the fact that there have in reality been few requests for help with financing from customers and suppliers. As the company moves from component sales to system sales, it is high time for a re-think! Instead of a sales interest of 10%, it would be possible to have a negative sales interest of a few per cent. The difference between selling with 3 months credit (which means keeping stock and paying interest) and receiving payment 3 months in advance (which means money in the bank and receiving interest) can be substantial. It could well mean an extra £3-4 million annually.

Adding all this together, we could be talking about a figure in the order of £6 million, although things in real life may be more complex than just adding these numbers together.

However, it could be realistic to talk of a yearly business development potential of some £3 million. To put this in perspective, OPERATOR's last annual result after finance costs was a profit of £0.9 million.

With that type of pay-back potential, it ought to be possible to spend substantial resources on a Business Engineering Programme.

## *The recruitment of a Business Engineering leader*

As we have seen, it will be helpful if there is a person devoted to leading the business development programme. Most likely this leader will be recruited internally.

> The choice in OPERATOR was between Leonard Hewson and Carol Parker. Leonard was considered to be too grumpy and Carol too easily offended. So the Managing Director decided to handle it himself.
>
> 'I can hardly think of anything more important to do,' said Charles once he got wind of the potential of £3 million per year.

## *Recruitment of Business Engineers*

The Business Engineers will also mainly be recruited from existing staff, as the basic skills needed are technology and application experience. It takes many years in the industry to build up this experience.

## *The business development programme*

The business leader should run a programme consisting of three blocks:

**1. Business economics**

**2. Business communication**

**3. Business strategy**

Since it is an overall cultural evolution towards business awareness that we desire, it is natural that the programme contains a large measure of training — but that is not enough. Many policy decisions will have to be reconsidered and both the organisation and its routines will have to be re-appraised.

I will comment on each of these three blocks in turn but I want to stress that every business development programme is unique because every company is unique.

The main ingredients will always be there but the emphasis and the details are different. There is, for example, good reason to spend time and effort on penetrating the use of working capital in OPERATOR, whereas another company may already have solved this problem by asking customers for substantial pre-payments.

Before kicking off the programme and involving many of the company's key people, a pre-study should have been made. It is when key people get involved that it starts to cost money, since their alternative cost is high.

## *1. Business economics*

### Why everybody needs to know business economics

We have seen earlier that the post-industrial company has to be run on the basis of 'management by objectives'. As most of the important objectives and restrictions are expressed in economic terms, the key people have to understand the economics of business development.

### Never try to compete with the finance director

Let us not over-emphasise these economics studies, however. The aim is not for the business people to compete with the specialists in the finance department, but to have an awareness of the financial implications of all aspects of the business activities.

To become a Chartered Accountant takes many years of study and long experience. Our studies of business economics should be focused on the areas a business person can influence and will by-pass many of the difficult areas such as financing, taxation and accountancy.

### Separate business from property and financing

The following objective is taken from a consultancy firm:

'Our objective is to have a net margin after depreciation and finance costs of 10% at 7% inflation, and our properties should give a direct return of at least 5%.'

In order to understand an objective such as this and to translate it into concrete action, it is necessary to have a very good knowledge of both economics and property management.

Additionally, the objective seems to indicate that profitability itself is the main goal rather than a constraint along the path to the goal of the business.

A good way of making profitability requirements more manageable is to separate the business activities from property and financing. Property should be organised in a separate business unit, effectively renting out space to the other business units at market rates. This will give us a good indication of the profitability of our property.

The finance function can also form a business unit itself, charging other business units — including the property unit — an average alternative interest. This makes it possible to measure efficiency in our finance function too.

## A sensible approach

One trick to make profitability targets manageable is to use one of the measurements we defined in Chapter 3; profitability measured as ROCE (return on capital employed) or, in some cases, profitability per head.

When taking this approach, it is important to use the alternative interest as the financial cost, not the real interest, which is something a business person cannot influence and which will fluctuate during the course of the year.

A profitability concept that is correctly defined means that we will have struck the right balance between business economics and finance. This way, business people do not need to get too involved in financing, taxation, accountancy etc. These questions should be left to the specialists, the finance director and his or her department.

The business people can now instead concentrate on business decisions such as:

- 'Are we in a position to try to increase market share, or should we concentrate on making more money from the market share we hold already?'
- 'Is it sensible to give our agent two months extra credit?'
- 'How do we justify extra advertising?'
- 'Should we employ another business engineer?'

## The finance director is not responsible for business policy

In a post-industrial company, the finance director should not be responsible for making economic decisions. He or she is, however, responsible for the system used to track economic decisions.

I sometimes compare the co-operation to that between a Formula 1 racing driver and his chief mechanic. It is natural that the mechanic understands the steering system better than the driver, but no-one suggests that he should drive the car in the race.

I would therefore contend that, it is possible to train in business economics people who do not have a previous or formal financial background, provided we concentrate on business-related issues.

## Practice makes perfect

Once the key people in the company have acquired a common frame of reference in business economics, it is time to put it into practice. We do this by starting one or more projects.

For the company to practise pricing according to customer-perceived value, it is necessary to analyse carefully our competitive features and the advantages we can offer to our customers. Systems for price differentiation will have to be drawn up and so on.

In order to apply market-based and capacity-oriented business selection, our calculation and accountancy methods must be reviewed, any bottlenecks in production need to be analysed, possibilities for incremental business must be examined etc.

## 2. Business communication

In traditional industrial (mass-) marketing strategy, the competitive weapon of promotion simply meant aiming to sell more of an already well-defined product. In post-industrial business development, our investments in personal sales and advertising are directed more towards explaining the benefits of what we are offering.

Market-based pricing is based on an assessment of the value perceived by the customer. The value has to be explained fully to the customer or else he will not perceive it, which will reduce the base for our pricing.

We must be able to answer the customer's justifiable question: 'Why should I pay you this much?' When we run out of answers, the price will be squeezed down.

As it is likely that most post-industrial business deals are made in face-to-face situations with customers, it is important for our business engineers to be good communicators and have the appropriate support.

In order to communicate our business, we need to improve the communication skills of our Business Engineers by a mixture of project work and training. First we should review the basic arguments for our business and make sure that we put them across effectively in mass communication and sales support material, such as brochures, overhead transparencies etc. Thereafter we should practise personal presentations in front of an audience.

It is then time to look at the way in which we write quotations and offers. Many quotations today are largely legal documents which border on 'sales discouragement'. They clearly spell out how much everything costs and stress the areas where we are **not** to be held responsible, but they tend to say very little about the customer's benefits.

## Business training

Since post-industrial business deals are struck individually with each customer, often in an international environment, a traditional sales training is probably insufficient.

It is no longer a matter of 'a thousand ways to close a deal'; it is all about doing good business.

In a post-industrial business negotiation, many factors will be unknown so the business awareness or acumen of the Business Engineer will be put to the test.

The Business Engineer has to possess a sixth sense to be aware of the customer's real needs. He or she has then to assess quickly and accurately which of our products or offerings will fulfil those needs as fully as possible and how to present it. The offer must be on terms that are acceptable to our own company, with due consideration of our current capacity situation, and should be put forward in such a way that the customer will perceive maximum benefit.

To master this difficult art perfectly is perhaps impossible; certainly it will need to be practised and improved throughout the active business career.

## 3. Business strategy

You might well ask why we did not start by defining the overall situation and working out a busines strategy.

The reasons are many. To avoid the strategy development degenerating into an abstract exercise, it has to be expressed in terms of money. This can only be achieved if a common frame of reference of business economics already exists.

The most important strategic decisions have been taken in the first stage, the business economics. These are market-based pricing and market-based business selection.

The strategic decision to implement a business-oriented company culture has in effect already been taken once we decide to start the post-industrial business development process.

Business strategy discussions can wait — but not indefinitely.

In post-industrial business development, we aim to become world leaders in a niche market. This means that we need frequent strategic check-ups, particularly looking at market segmentation and resources.

How strategy development should be carried out has been shown in Chapter 12.

# Summary

It is a major decision to leave behind the old industry strategy of cost-based pricing and cost-based business selection. The old strategy was almost self-governing. You could even say that it was possible for a college graduate in economics to take the most important business decisions in the company; namely, setting the prices and selecting business.

He or she hardly needed to know anything about the business itself. When optimising competitiveness through price, we calculated the lowest price on which we could survive — customers who accepted this price could do business with us. Both of these decisions could be computerised.

When changing strategy to market-based pricing and market-based business selection, completely different demands are placed on the organisation.

To set the price according to customer-perceived value (CPV), it is not sufficient to know our own offer in detail; we also need to understand the offerings from our competitors. This will help us define our UBRs ('unique buying reasons', or USPs — 'unique selling points'). These UBRs form the basis of any price differences between us and our competitors.

But even this is not enough. In order to assess our UBRs in the form of value to the customer, it is necessary to understand the customer's applications in enough detail to be able to explain the advantages and benefits of our offer, in terms they will understand.

Even knowing the price we can obtain does not mean that we will be prepared automatically to do business. We have to evaluate every business opportunity based on our company strategy and our current capacity utilisation.

To be able to apply market-based pricing and business selection you have to know the company's business well. Management by objectives and sound business awareness must now permeate the whole organisation.

Based on all this, the most important task for senior management is to recruit a business development leader and Business Engineers, and to develop an organisation giving them the right environment in which to work.

# Chapter 14

# A summary of the summaries

*New times, new heroes*

We are in a transitional period between the industrial and the post-industrial economy. Old recipes for success when marketing mass-produced, material products do not automatically work when we market products or systems whose value is largely their quality and knowledge or information content.

In the old industrial economy, there were some fundamental tenets: prices should be calculated; every business deal should cover its own costs; salespeople should work on a commission etc. In post-industrial business the opposite is often true.

In an economy where products and services become less and less tangible (in a material sense) and their lifespans become shorter, a new business method — that I call Business Engineering — is needed. One important ingredient is pricing according to customer-perceived value (CPV).

For this form of pricing to be successful, you not only have to understand your own products but also those of your competitors. Equally you have to understand each of your customers' applications and how your product can benefit them. On top of this you have to know **how to do business**. Individuals who are able to achieve all this can be called BUSINESS ENGINEERS, irrespective of their formal training. They are worth their weight in gold to a technology- and R&D-intensive company.

**One of the most important tasks for senior management in a post-industrial company is to fill the organisation with Business Engineers and create the right environment for them to work in.**

## *The goal — to what are we aspiring?*

Business engineers are not controlled by the whip or solely by the carrot — they will be motivated by goal orientation and management by objectives (MbO). The objective common to all parties involved with the company is **satisfactory growth**.

Ambitious employees prefer expanding companies with growing opportunities for interesting jobs. The owners want around twenty per cent as the return on a growing capital base. Customers, suppliers and society also all want to see a successful company grow.

## *The importance of a healthy relationship with the bank*

For growth to be considered satisfactory, it has to be balanced, it has to show long-term profitability and, lastly, it has to be at or above the market's own rate of growth (that is to say, market share must be increased).

Post-industrial companies must be strong financially, since the visible profitability will fluctuate with the economic cycle, due to high capacity costs and the large investments that have to be made in the development of products and markets.

These investments in service, product and market development tend to be written off in one year, even if they have a value that lasts for years. No banker dares to finance a substantial part of such investments. It is a lot easier to release further loans for a building project when you see it rise floor-by-floor than for a complicated computer program where it is hard to judge development progress.

It is therefore important to have a good equity ratio in order to avoid having to 'destroy' capital by halting important development projects or laying off valuable employees when a recession starts to bite. It is as important to delegate responsibility for the balance sheet as for the profit-and-loss statement. Profitability analysis must be forward-looking and viewed over a full business cycle.

It is usually necessary to have a substantial share of a well-defined market segment in order to achieve satisfactory long-term profitability. The growth therefore has to be faster than the market's growth.

Therefore it is essential that post-industrial business planning starts by examining the 'red/green position' of the company. Only if the company can show good financial strength, in the form of its equity ratio, and the outlook for long-term profitability is good (a 'green light' situation) will we be allowed to adopt an aggressive plan to increase market share. If we are 'in the red' (which could mean a strong profit-and-loss statement but a weak balance sheet) our main objective for the planning period will be to make enough money to restore our financial strength.

## Follow the 'straight and narrow'

When we are given the green light to plan for expansion, it is important to follow the straight and narrow road towards satisfactory growth.

The aim is to dominate a specific market, irrespective of whether or not we are going to compete overseas. Post-industrial companies always have international competitors, even in their home markets.

The more narrowly defined the niche, the easier it will be to dominate and defend.

## A crossroads en route

Having travelled some way along our desired route, we come to a crossroads: 'Do we adopt fully the principle of market-based pricing, or do we stick to the tried and tested way of setting prices by calculating our costs?'

Provided we are selling simple products on price, and as long as costs are easy to adjust, we should retain cost-based pricing. But when we market sophisticated products and/or services, we have to optimise our development potential by charging prices according to customer-perceived value.

As a result of our choice of market-based prices, we will also opt automatically for a method of selecting business based not on cost, but on an optimum balance between price and capacity utilisation.

## Post-industrial business selection

A post-industrial company cannot use full-cost calculations or contribution calculations to prioritise business deals. What is required is a consistent 'opportunity cost' (or 'alternative cost') philosophy. We can identify six different situations:

1. Short-term decisions where there is overcapacity;
2. Short-term decisions where there is undercapacity;
3. Long-term decisions with undercapacity and a red light;
4. Long-term decisions with undercapacity and a green light;
5. Long-term decisions with overcapacity and a red light;
6. Long-term decisions with overcapacity and a green light.

In all these situations we have to apply a different view of what is 'profitable'.

We have now reached an important stage where we have a common frame of reference for post-industrial business development. We can now begin to expend our energy on developing our competitiveness.

## System development

In the industrial society the emphasis was on developing **products**. In the post-industrial or information society this becomes **systems** which comprise three components equivalent to what the computer industry calls hardware, software and 'documentware'.

The aim has always been to create a benefit that the customer is prepared to pay for. The challenge now is to get paid for something that consists of more 'knowledge' than 'mass'.

The possibilities of getting paid for the benefits and values we create have to be built into the product early in the development phase; the communication of customer value has to be developed alongside the technology itself. There are some sophisticated values that are impossible to communicate or get paid for. However, a small change in the technical construction of the product or system can often resolve such a problem — but not after the system has been launched and sold.

## Price development

In the industrial society companies were 'hit by price development', meaning that prices were inexorably driven down by strong price pressure from customers and competitors. In the post-industrial economy companies develop their own prices as an integral part of their business strategy.

The words 'market price' mean 'the price the market is willing to pay'. When the product is highly standardised and we can adjust our costs to match the demand, this price is where supply and demand curves intersect.

**But when we market sophisticated systems, the market price is whatever we can make it.**

Once costs have been discarded as a basis for pricing, we soon find that the price we can get is the price we can justify; simply having high costs is no justification for high prices.

The price is therefore related to how we communicate the value of our offering to the customer. If we can only explain 30% of the advantages of our offer, we can only expect to get paid half as much as if we could explain 60% of the advantages. Very few will succeed in communicating 100%. The only person who can make us come closer to communicating 100% of the value of our offering to the customer is the Business Engineer.

In post-industrial business development it is not enough to develop a single price. Not all customers will obtain the same benefit from our products; not all should therefore necessarily pay the same price. If this is to be acceptable in the marketplace, we have to apply a well thought-through price differentiation.

Once we have accepted the principle of pricing according to customer-perceived value, we are committing ourselves to a revolution in our business thinking. For example, we cannot sell whilst expecting the customer to pay the freight and absorb any currency fluctuations; the customer's perceived value does not increase with distance.

## Distribution development

When selling basic, undifferentiated products, it is possible to export extensively by 'skimming the cream' off many overseas markets with limited marketing investments, for example by using commissioned agents.

The situation becomes quite different when marketing sophisticated systems. Normally this means large local investments in customisation of products, as well as local and central service and marketing. The ambitions have to be correspondingly higher when entering a new geographical market.

Instead of 'trickle exports' we should practise 'multi-domestic business development', which means our company should be perceived as 'local' in each of the markets we have chosen to develop.

A less ambitious local representation than a subsidiary company can only be justified if we limit our offering to components. Franchising and licensing can be interesting alternatives, particularly when marketing products or services that are almost pure knowledge.

## *Communication development*

The post-industrial company, naturally, is operating in the post-industrial society, also known as the information society. The skill the company displays in communication determines its future. Both the product/system and the price have to be communicated.

It is not enough, however, only to communicate with customers. As the post-industrial company largely depends on the service content of its products, communication has to start with the personnel. It is their acceptance of goals and methods, and their creative participation, that form the basis for the identification and communication of customer benefits.

To avoid worrying shareholders to the extent that they will withdraw their capital when profitability fluctuates, we have to keep them informed about the way our business develops. We must also aim to have the best possible contacts with other interested parties.

Since the communication the company maintains with its interested parties is so vital, it should not be left to enthusiastic amateurs. You need more than the latest word processor to write excellent scripts. There will be times when the company needs communication professionals.

## Strategy development

In post-industrial business development we aim to become world leaders in our niche area. It is important to have frequent checks on market segmentation and the use of resources.

Strategic planning work must not be considered so important that it leads to 'paralysis by analysis'. We can avoid this by an incremental planning process (i.e. going round the planning loop several times), basing the first round on hypotheses. When the plan is ready we can test those hypotheses that are crucial to our success, and spend far less time on those that are of less importance.

Our financial situation forms the very foundation on which the plan rests. Crudely, it can decide whether we must concentrate on saving the company from bankruptcy, or if we are in a position to capture market share. Thereafter the main questions are how we define and prioritise our business.

## Organisational development

It is not the organisational structure that decides the success of the post-industrial company, but the interaction between the Business Engineers and the rest of the management. This interaction has to be flexible rather than locked into any rigid organisational pattern.

The difference between the old, familiar ways of doing business and the new method I am calling Business Engineering is of such a magnitude that it often requires an 'evolution leader' to carry through an 'evolution programme'.

The old 'administrative' company culture, based on calculation and control, must be replaced by an organisation which thinks and breathes business. It may well cost time and effort to achieve this, but the resources that will be released for development will be large — often 5 to 10 per cent of turnover.

# Appendix

# OPERATOR Ltd.

## ANNUAL REPORT, 1994

# Financial summary

|                                    |     | 1990 | 1991 | 1992 | 1993 | 1994 |
|------------------------------------|-----|------|------|------|------|------|
| **Sales**                          | £M  | 8.1  | 24.2 | 34.1 | 4.1  | 33.2 |
| **Profit before provisions & tax** | £M  | −1.4 | −0.7 | +2.1 | +1.5 | +1.0 |
| **Average number of employees**    |     | 246  | 212  | 214  | 218  | 217  |
| **Investments in machines & equipment** | £M | 1.0 | 2.1 | 2.3 | 1.5 | 0.6 |

# Directors' statement

## Product range

Microprocessor-controlled systems, and sophisticated programming/analytical services, for Total Quality Management and reduction of energy consumption in combustion processes.

## Market development

The considerable interest in our flame control systems and consultancy services remains unchanged. The rate of investment in the industry had been declining over the last few years, but stabilised in 1994. Despite a difficult business climate we have managed to maintain our sales at a constant level.

## Investments

Following substantial investments over the past three years, investment in machines fell last year to £0.6 million. This, however, does not mean that our investments have been reduced in real terms.

1994 was a year of hectic product and market development. If these investments, very important to the future of the company, had been treated in the same way as machine investments, operating profit would have been substantially higher. We have, however, adhered to prudent accounting rules and capitalised only a small part of these expenses.

## New subsidiary

To secure a foothold in the world's fastest growing economic sector, namely the ASEAN countries, we have established our own sales office in Singapore — OPERATOR (Far East) Ltd.

This means that we now have wholly-owned subsidiaries in Belgium, Germany, Scandinavia, the United States and south-east Asia.

## Prognosis for 1995

Although a general improvement in the business climate in our trade is forecast for 1995, it will hardly affect OPERATOR Ltd. Our business is situated late in the economic cycle since we market products for services and investments.

The investments made in market and product development in 1994 should show results in 1995. It is not possible to predict accurately the effect that they will have on the operating profit, but it is highly likely that it will exceed 1994 levels.

# OPERATOR Ltd.

## PROFIT-AND-LOSS STATEMENT, 1994

|  |  | £ M |
|---|---|---|
| 1. | NET SALES | + 33.2 |
| 2. | COST OF SALES | − <u>30.5</u> |
| 3. | RESULT BEFORE DEPRECIATION | +2.7 |
| 4. | DEPRECIATION | − <u>0.9</u> |
| 5. | RESULT AFTER DEPRECIATION | + 1.8 |
| 6. | FINANCIAL REVENUE | + 0.1 |
| 7. | FINANCIAL COSTS | − <u>1.0</u> |
| 8. | NET OPERATING PROFIT | + 0.9 |
| 9. | EXTRAORDINARY INCOME | + <u>0.1</u> |
| 10. | PROFIT BEFORE PROVISIONS & TAX | + 1.0 |
| 11. | PROVISIONS | − 0.2 |
| 12. | TAX | − <u>0.4</u> |
| 13. | NET PROFIT | + 0.4 |

# OPERATOR Ltd.

## BALANCE SHEET, 1994
## as at December 31, 1994

| ASSETS | £ M | EQUITY & LIABILITIES | £ M |
|---|---|---|---|
| 1. Land | 0.2 | 11. Shareholders equity | 2.8 |
| 2. Buildings | 1.5 | 12. Retained earnings | 0.4 |
| 3. Machines | 1.5 | 13. **Total Equity** | 3.2 |
| 4. Long-term | 0.3 | | |
| 5. **Total Fixed Assets** | 3.5 | | |
| | | 14. Long-term debts | 10.8 |
| 6. Stock & work-in-progress | 8.5 | 15. Deferred taxation | 1.8 |
| 7. Accounts receivable | 7.6 | 16. Overdraft | 1.0 |
| 8. Cash | 0.9 | 17. **Total Long-term Liabilities** | 13.6 |
| 9. **Total Current Assets** | 17.0 | | |
| | | 18. Advance payments | 0.1 |
| | | 19. Accounts payable | 2.2 |
| | | 20. Other short-term liabilities | 1.4 |
| | | 21. **Total Short-term Liabilities** | 3.7 |
| 10. **Total Assets** | 20.5 | 22. **Total Liabilities & Equity** | 20.5 |

## Report of the Auditors

We have audited the accounts in accordance with auditing standards. In our opinion the accounts give a true and fair view of the state of the company as at 31 December 1994, and of the profit and source and application of funds of the company for the year then ended and have been properly prepared in accordance with the Companies Act 1985.

**Williams, Livingstone and Grant**

Chartered Accountants

## Join the International
## BUSINESS ENGINEERING CLUB!

The Business Engineering concept is under constant development by Teknosell® and the Business Engineers practising it in the marketplace.

In order to facilitate the exchange of ideas and experiences we have started **The International BUSINESS ENGINEERING CLUB,** which now has approximately 500 members — a number that is quickly increasing.

To stay informed by being registered as a Club Member is free of charge. The events are offered at cost.

Are you interested in contributing and developing your own skills in Business Engineering? If so, please register your name and address with:

Björn Rosvall, Chairman of the International
**BUSINESS ENGINEERING CLUB**
**15, Westwell Road**
**LONDON SW16**
**Fax ++44 (0)181 679 62 92**
**e-mail: Rosvall@ihm.se**

# Glossary

The meanings attached to the terms that follow are not meant to be rigorous definitions but rather working explanations to help the reader in understanding the economic and business concepts introduced in this book.

| | |
|---|---|
| alternative cost | the cost difference between the alternative under scrutiny and the best-known benchmark or alternative |
| alternative interest | the interest the owners know they can obtain through an alternative use of the capital (e.g. depositing it in the bank) |
| business area | an additional geographical level within a *business division* |
| business division | a strategic unit within a *company* which can be assigned a reasonably independent balance sheet and profit-and-loss account, probably controlling resources to pursue a distinct business idea |
| business unit | a combination of a *business division* and a *business area* |
| (to) capitalise an investment | to list an investment on the balance sheet |
| CIF | 'cost including insurance and freight' |

| | |
|---|---|
| company | a strategic unit with an independent balance sheet and profit-and-loss account |
| concern | a strategic unit with independent financing |
| contribution | the additional income from a business deal, having deducted *variable capacity costs* (only those costs directly attributable to the deal, i.e. costs that would disappear if the deal was not done) |
| contribution ratio | (1) *contribution* as a percentage of sales value; (2) the average contribution needed to cover *fixed capacity costs* through the year and the profitability requirement |
| contribution$_2$ | a hybrid between *contribution* and *full cost* calculations — the higher the index the more costs are included |
| current ratio | the ratio of current assets to short-term liabilities |
| DDP | 'delivered duty paid' |
| equity | risk capital |
| equity ratio | risk capital expressed as a proportion of total assets |
| financial strength | see *equity ratio* |
| fixed capacity cost | invariable overheads |
| FOB | 'free on board' |
| free loans | monies received in advance of liabilities (e.g. payment in advance, VAT which is received but not yet returned) |
| full cost (calculation) | all costs, not only those attributable to the particular business deal, therefore including overheads (can include *alternative interest*) |

| | |
|---|---|
| internal interest | see *alternative interest* |
| liquidity | funds available for immediate payments |
| opportunity cost | see *alternative cost* |
| opportunity marginal calculation | the difference between alternative deals, expressed in terms of their contributions |
| payback time | investment divided by income surplus per month (etc.) |
| profitability per capita | [total income — total costs] / number of (strategic) people |
| return on capital employed | [total income — total costs] / [total assets - free loans]; note that total costs include depreciation but exclude financial costs |
| return on equity | net profit after costs (and after tax), expressed in relation to total equity |
| ROCE | see *return on capital employed* |
| safety margin | difference between budgeted sales and those sales needed to break even (i.e. to cover fixed capacity costs in the year and profitability requirements) |
| sales interest | the alternative cost of the working capital; generally, it is the net profit margin needed to cover return of the additional capital used in the project |
| variable capacity cost | variable overheads |

# Index